The Air Campaign

PLANNING FOR COMBAT

Col. John A. Warden III, USAF

Introduction by Gen. Charles L. Donnelly, Jr., USAF

toExcel

San Jose   New York   Lincoln   Shanghai

The Air Campaign

This edition published by toExcel Press,
an imprint of iUniverse.com, Inc.

For information address:
iUniverse.com, Inc.
620 North 48th Street
Suite 201
Lincoln, NE 68504-3467
www.iuniverse.com

Opinions, conclusions, and recommendations expressed or implied within are solely those of the author and do not necessarily represent the views of the National Defense University, the Department of Defense, or any other U.S. government agency.

Photos credited to the USAF Photographic Collection, National Archives.

ISBN: 1-58348-100-1

Library of Congress Card Catalogue Number: 98-88494

Printed in the United States of America

To my wife, Margie

Contents

Illustrations

PHOTOGRAPHS

TABLES

Foreword

Since its publication in 1988, *The Air Campaign* has been translated into more than a half dozen languages, and is in use at military colleges around the world. I hope this new edition, and one of the first books ever to take advantage of on-demand publishing made possible by the Information Revolution and its marvelous offspring, the Internet, will be useful to a new generation of strategic thinkers whether they are concerned with military, political, or business competition.

Just three years after the first publication of *The Air Campaign*, I had an extraordinary opportunity given to few officers who have thought and written about war. On the 10th of August 1990, I presented to General Norman Schwarzkopf a proposal for an air campaign to reverse the Iraqi aggression on Kuwait consummated the previous week.

In the time between writing the book and its publishing, I had refined and extended many of the ideas first presented in *The Air Campaign*. The most significant of these refinements was the development of the theory of the enemy as a system— which flowed from the center of gravity concepts which were such an important part of the book.

By the time of the Iraqi invasion of Kuwait, my colleagues and I on the Air Staff had been able to apply and wargame the concept of the enemy as a system against the then all consuming scenario of a war with the Soviets in Central Europe. Thus, when the opportunity arose to make an encompassing proposal to General Schwarzkopf, we were able within forty-eight hours of his request to build and present to him an outline for a strategic and operational level plan. He found the plan quite attractive as did General Colin Powell, then Chairman of the Joint Chiefs of Staff, when we presented it to him the next day, at General Schwarzkopf's direction.[1] (See the Epilogue for more detail on how we used these and other ideas in the war against Iraq a few months later.)

On rereading *The Air Campaign* some thirteen years after I wrote it, I find that I am still comfortable with most of the observations and prescriptions. I would, however, make a few changes in wording and emphasis. First, I would emphasize much more emphatically the importance of thinking strategically and of concentrating efforts against the strategic centers of gravity. Doing so, especially if you have available precision weapons and stealth technology, allows far more rapid and more economical attainment of objectives than focusing efforts on tactical or even operational level targets. In this light, I would no longer equate "distant interdiction" with strategic attack. I would also make clearer the idea that it is entirely possible to win major competitions without ever dealing with the opponent's fielded forces.

I have found the concept of the enemy as a system and the ideas originally developed in *The Air Campaign* to have equal relevance to business, political, and military competition— because above a tactical level, all competition is remarkably similar. The good ideas that apply to one sphere apply equally to the others. Readers of this new edition who are not military professionals should keep this in mind as they read *The Air*

---

1 A number of books have talked about this meeting with General Schwarzkopf. See for example, General Schwarzkopf's book It Doesn't Take a Hero, General Colin Powell's book, My American Journey, Rick Atkinson's book Crusade, Michael Gordon and Bernard Trainor's, The General's War, and Rich Reynolds' book, Heart of the Storm.

Campaign for they will find that they can apply many of the principles to their own business. For those who would like to see these and other ideas more explicitly developed for the business world, I am currently co-authoring a new book tentatively titled *The Prometheus Principles: Leadership in the Digital Age*, which should be out in the summer of 1999. I also intend to write a book about the Gulf War and its aftermath, but that is probably a couple years away.

It seems to me that there is one crystal clear idea in this fast moving world in which we find ourselves; the only way to true success and sustained differentiation is through better thinking, strategy, and planning. Everything else can be quickly copied and used against you. It is my hope that everyone reading this book will walk away with concepts which will help them in their areas of competition. May all your campaigns lead to success!

Preface

The Air Campaign is an attempt to come to grips with the very complex philosophy and theory associated with air war at the operational level. This book is for combat officers of any Service who might find themselves on an operational-level staff. More specifically, it is for the air officer who wants to think about air campaigns before called on to command or staff one. It is devoted to how and why air power can be used to attain the military objectives needed to win a war.

What this book is *not* is quite important also.

This book is not about tactics and does not address how to bomb a target. It is not technical and does not address specific weapon systems. It is not specific to any particular air force and thus does not address directly any of the various disputes over doctrine that are common in many air forces. Likewise, it avoids using terms that recently have come into vogue but are still too esoteric to be widely understood or usable.

As a consequence, older words like "front" are used, rather than more specific ones such as "forward line of troops." My belief is that the more general term better conveys the image needed for a conceptual discussion at the operational level.

Two other areas not addressed in this book are the uses of space and nuclear weapons.

With respect to the use of nuclear weapons, one either believes that their use cannot be squared with any rational view of war, or one believes that they are in some cases usable in consonance with traditional ideas. I have not discussed space operations primarily because the operational-level commander at present has no direct control of space assets. In the near future, man certainly will spread out through the solar system. If war goes with him, the principles should not change significantly, although the concepts of depth and time may become more important than ever before.

Technologies change with great rapidity; consequently, any book on air warfare that went into depth on particular technologies would become dated very quickly. I believe that the operational-level commander must first master the basic philosophy and principles of warfare. Only then can he make current or new technologies his servant. If he tries to reverse the process, he will be unable to set a course and will be driven haphazardly by every change in the storms of technical development.

On the horizon at the end of the 1980s are exciting possibilities for directed-energy weapons and for short- to medium-range ballistic missiles armed with conventional ordnance. Either or both—or something as yet undreamed—may become quite important, but only because they allow greater concentration of power or increased mobility. Successful employment will depend on using new systems in consonance with principles outlined in this book.

The reader also will note that this book includes little discussion of aircraft carrier-based air power. The lack of discussion was not meant to denigrate carrier air power by any means; indeed, in any conceivable major war fought by the United States, aircraft carriers will be a necessary part of the offensive needed to win the war. However, since this book is meant to be a guide to the *use* of air power, as opposed to a history of it, examples from land-based air seem sufficient to illustrate my observations.

The theory at the operational level should be the same, regardless of the point from which aircraft or missiles launch.

As the time since the last war lengthens, military institutions tend to focus increasingly on future strategies and the force structure needed to support them. Such a focus is necessary, but plans for fighting a future war with future force structure should not be confused with plans for fighting a war that might start tomorrow. For the latter, only existing forces are of significance, although campaign plans for long wars could take into account new equipment and new units that may materialize as the result of national mobilization. In essence, however, operational-level theory is not concerned with developing future force structure: It is quintessentially concerned with using what is available.

The Air Campaign is, very simply, a philosophical and theoretical framework for conceptualizing, planning, and executing an air campaign. To the extent that it assists any planners in arranging their thoughts—before they are in the thick of battle—it will have achieved its ends.

Acknowledgments

My special thanks to Colonel Roger P. Fox, US Air Force, my military history professor at the Air Force Academy, and to Dr. Frederick H. Hartmann, my grand strategy professor at Texas Tech University.

Thanks also to Colonel Michael D. Krause, US Army, my research paper sponsor while I was at the National War College, for his encouragement and help with *The Air Campaign*.

Introduction

By Charles L. Donnelly, Jr.

This book is the start of something very important—it integrates historical experience into a clear, visionary set of conclusions and guidelines for using air forces to achieve strategic goals in a war. This book is exceptional, because it is the first book that thoroughly covers the area between the selection of national objectives and tactical execution at the wing and squadron operations levels.

A book of this type has been needed for a long time.

Centuries of land warfare passed before writers like Sun Tzu succeeded in capturing for readers the essence of success in war. We are fortunate, only decades after airplanes first were used in combat, to have a coherent synthesis of historical experience that is both consistent with history and prescriptively useful for future employment of air power.

Colonel Warden does not write about tactics or specific weapons, nor does he rely on a specific war. But he does tell us how to use air power most effectively. It is a book about art—operational art—as it should be practiced by an air component commander, and it ties directly to the enduring principles of war.

The principles of war, so eloquently discussed by Carl von Clausewitz in his famed book, *On War*, are the same for both

---

General Charles L. Donnelly, Jr (USAF-Ret.), was Commander in Chief of US Air Forces in Europe from November 1984 to May 1987 army and air forces. Why, then, does an air force need a special analysis? What will be reaffirmed after reading *The Air Campaign* is that the speed and range of air forces pose special problems and offer special advantages that center around the principles of mass and concentration and their corollary, economy of force. Colonel Warden's book provides insightful discussion of such crucial topics as the commander's choice of an offensive or defensive orientation, the requirement for air superiority, the influence of ground-based defenses, and the intriguing new idea of air reserves.

The Air Campaign makes it clear that the air commander frequently will have a choice between offense and defense. An air force on the defensive faces greater risk to itself and the total war effort than an air force on the offensive. A special feature throughout this offense versus defense discussion is an enumeration of the best choices in all the warfighting situations in which air commanders could find themselves—inferior forces, superior forces, guerrilla war, theater war, and even the danger of a massive enemy breakthrough on the ground. Also answered is the question of whether air force fighter defenses are better when spread to defend large areas or when concentrated, leaving some areas less well defended.

Whatever the choices for offense or defense, the air campaign cannot succeed until air superiority is achieved.

We consistently tout air superiority as our number one priority. While this thought seems logical to airmen who live and breathe air power, it sometimes is difficult to convince others of this fundamental doctrine. This book offers a convincing argument on the need for air superiority. It also provides a balanced analysis of air power's other roles in supporting the joint force commander—including taking the sting out of opposing air defenses, ground or air.

Enemy ground-based air defenses are targets that will be defeated at times and places of our choosing. Any ground-based air defense system has vulnerabilities that reduce its strength. For example, it is never equally strong throughout its length and breadth, it has flanks, it is immobile compared with air power, and it is normally oriented toward a specific threat. These vulnerabilities can be exploited in a well-planned air campaign. And because the vulnerabilities are not technological, but inherent in the concept, a ground-based system never will be able to stand alone against the unpredictable shock and violence of concentrated air attacks.

One way to increase the concentration of air attacks against any set of targets is to retain some air power to meet the unexpected—whether providential or disastrous. Errors are made on both sides in war, and reserves of air power permit exploitation of enemy errors, or they can be directed into the breach against an enemy attempt to exploit our mistakes. This concept of an air reserve rounds out this thoughtful book of operational-level art, providing the reader with a final technique to plan air campaigns against a wide range of enemy capabilities.

After finishing this book, the reader will realize that planning an air campaign is an art, because the fundamental decisions that will win or lose an air campaign depend on the logical thinking of the human computer. Detailed operations research techniques and probabilities are necessary inputs to our planning process. But Colonel Warden will lead you back, recharging your belief in fundamental logical tools that can be applied at first by skilled airmen, with sophisticated math taking a proper role of backup and verification.

It is possible for an air force to have absolutely superior forces—numerically and qualitatively—and lose not only the air war but the entire war. I strongly recommend *The Air Campaign*, because it provides the air commander the intellectual wherewithal needed not only to avoid losing, but to win.

The Air Campaign in Prospect

Thinking about war and how to fight it is an immensely difficult undertaking, because war is the most complex of human endeavors. Because it is so complex, it must be broken into component parts that can be examined, studied, and used. Clearly, war can be broken down in a number of ways.

Wars are big and small, limited or unlimited, nuclear or nonnuclear, geographically confined or worldwide. Although these divisions are useful, they do not give a very good basis for planning or directing operations, because they are still too encompassing. From this standpoint, it becomes more useful to break war down into parts that are related to decreasing levels of responsibility.

THE LEVELS OF WAR

The four levels of war discussed here include *grand strategic, strategic, operational,* and *tactical.*

The *grand strategic level* of war is the place where the most basic but most consequential decisions are made. Here, a country determines whether it will participate in a war, who its allies and enemies will be, and what it wants for the peace. In World War II, President Franklin D. Roosevelt, British Prime Minister Winston S. Churchill, and Soviet Premier Josef Stalin were arbiters of the grand strategy that largely dictated the military strategy of the states opposing the German-Italian-Japanese Axis.

The *strategic level* of war concerns the overall conduct of the war, the approximate forces that will be made available, and the weights of effort in various theaters. To illustrate, the decision to emphasize Europe, rather than the Pacific, in World War II was a strategic decision. Similarly, the decision that Germany would be defeated by land invasion, rather than by blockade or air attack, was strategic. General George C. Marshall, US Army Chief of Staff, and British Field Marshal Sir Alan Francis Brooke were chief architects of American and British stategy in this conflict. Twenty years later, the decision to limit the number of men and aircraft available for use in Vietnam was a strategic decision, as was the decision to limit the air campaign over North Vietnam.

The *operational level* of war is the next level below strategic. It is primarily concerned with how to achieve the strategic ends of the war with the forces allotted. It is the level at which plans are made for the actual employment of land, sea, and air forces and the level where these forces are used in the course of a campaign. Generally, a theater commander is concerned with operations, as opposed to strategy. In this sense, General Dwight D. Eisenhower, General Douglas MacArthur, and Admiral Chester W. Nimitz were operational-level commanders (although there were strategic and even grand strategic implications in many of the plans they made and the operations they carried out).

The lowest level of war is the *tactical level*. This level is where opposing forces physically meet, where objectives are unambiguous—like taking a specific hill with a company, meeting and sinking an enemy ship, or fighting an aerial battle with an opposing fighter. The word "unambiguous" is important, because the men responsible for planning and carrying out tactical movements normally are told by higher authority precisely what they are supposed to do. This by no means suggests that carrying out tactical orders can be done without enormous mental effort; it merely means that the mental effort can be directed to a reasonably discrete objective, as opposed to the very complex objectives that must be selected and addressed at the operational and strategic level.

Many books have been written on the strategic level of war; indeed, one of the most famous and most useful is *On War*, written a century and a half ago by the Prussian strategist Carl von Clausewitz. Likewise, numerous books are available on the tactical level; in fact, the majority of war books are really at this level. Most works on submarine warfare, for example, or aerial combat, naval engagements, and infantry attack are concerned with the tactics of man against man.

Surprisingly—or perhaps not—almost nothing has been written since the immediate post-World War II period that deals with theory and practice at the operational level, especially for air warfare.[1] What explains the absence of works on the art of conducting an operational level campaign?

First, it is a difficult area to address. One can discuss strategy with broad gestures across a map of the world, and tactics are something with which many have had direct experience either in war or in training.

Second, after World War II, there was a certain sense that nuclear weapons had made the massing of armies, navies, and air forces obsolete. The near simultaneous rejection of history (especially in the United States) as the peacetime soldier's only window on war further pushed the operational level of war into the category of the arcane. When the last of the officers with high command or staff experience from World War II and Korea retired from active duty, a whole body of hard-won knowledge was lost.

Nevertheless, operational thinking remained essential, and genuine understanding of it remained vital. Many current problems over the uses of the various Armed Services stem from a lack of coherent doctrine on how they should be used individually and collectively in an operational campaign to secure some strategic end. This book is an attempt to fill that gap and to provide a framework for planning and executing air campaigns at the operational level.

In the belief that history is the only laboratory that we have in peacetime to develop and try theories of war, this book draws heavily on the last half century of air warfare. It uses examples from victor and vanquished alike—frequently, the better lessons are those learned in the aftermath of defeat. It tries to distill the lessons that can be drawn from many campaigns and many cultures. It does not suggest that a particular stratagem can be repeated in the future—although some can, given our short memories. Merely knowing that something worked once in the past may give a commander or planner an idea or the confidence to try a similar approach. How many victories—and defeats—came about because a commander had studied Hannibal's double envelopment at Cannae?[2]

Our focus will be on the employment of air forces at the operational level in a theater of war. Depending on the goals of the war, the theater may extend from the front to the enemy's heartland, as it did for the Western Allies after the Normandy invasion in World War II. Conversely, the theater may be a relatively isolated area, as in the desert war between Britain and the Axis in North Africa prior to November 1942.

In the former case, "strategic" air attacks on the enemy homeland affected operations throughout the theater and were of great interest to the operational commander. On the other hand, in the confined and isolated theater of North Africa, "strategic" attacks on German industry had little immediate local effect and were not of significant interest to the operational commanders on either side.

TWO LEVELS IN WESTERN EUROPE

In Western Europe, the strategic and operational levels nearly merged. The grand strategic goal was the unconditional surrender of Germany, and the military strategy chosen to realize the goal was ground penetration and occupation of the enemy state. Given that strategy, the operational commander, General Dwight D. Eisenhower, could not establish a bridgehead at Normandy and wait for the Germans to bleed themselves white trying to dislodge it. On the contrary, he had to move forward.

Necessarily, he had to use all available means to reduce the German capability to resist. Although not under his control, for a variety of reasons, the strategic air attacks on petroleum facilities, transportation networks, and power plants hundreds of miles behind the lines had at least as much to do with his eventual operational-level success as did the movement of armies on the ground.

Calling air attacks on the enemy heartland "strategic," as though they were on some special plane of their own, unrelated to the rest of the "real" war, can easily confuse us. In World War II, the Allied "strategic" bombing campaign had emasculated the *Luftwaffe* and forced its concentration in the defense of the homeland. Thus, the skies above Normandy during the Allied invasion in June 1944 were almost completely clear of German planes and the German army had the most extraordinary difficulties in moving forces to the front because Allied air forces made movement by day virtually impossible. No doubt the German operational commanders would have liked nothing more than to have the Allied air forces fighting desperately over Britain against a German "strategic" bombing campaign.

The key point in this example is that the strategic and operational levels merged. Operations from the lowest level to the highest are on a continuum and it serves us poorly to compartment them in such a way that we lose sight of their interrelations.

In North Africa, on the other hand, limited men and materiel and the general situation (Vichy France nominally controlling and thus protecting the Axis rear) made the concept of Allied operations relatively simple, at least until American entry into the theater in November 1942. Both sides wanted to destroy the other's ground forces; for the most part, they could only work against targets fairly close to the lines. (The exception was British interdiction of Axis shipping across the Mediterranean.) For the commanders on the ground, air attacks on Germany or Britain were of little immediate consequence, as they were unlikely to have any timely impact on their campaigns. Therefore, unlike General Eisenhower or German Field Marshal Karl von Rundstedt at Normandy, the North African com-

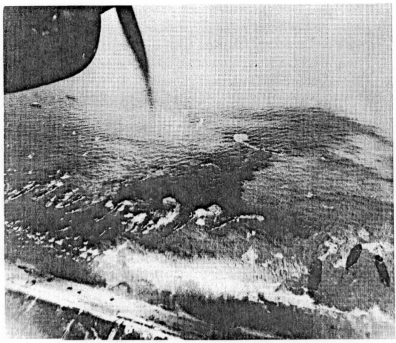

D-Day landing operations on the French coast on 6 June 1944

manders did not have to concern themselves with "strategic" air operations.

The theater commander should consider all kinds of operations that might have an influence on the campaign. If the will of the enemy people is vulnerable, the theater commander may want to concentrate efforts against that target. If the enemy is dependent on external supply, then some point in the supply chain may be the key to success. If the enemy is dependent on petroleum, then destroying petroleum networks may be the smartest move. In many cases, however, and especially against a modern, highly resilient industrial power, there may be no single key; thus, attacking a number of targets may be necessary—but targets carefully chosen to affect the enemy's center of gravity.

"CENTER OF GRAVITY"

The term "center of gravity" is quite useful in planning war operations, for it describes that point where the enemy is most vulnerable and the point where an attack will have the best chance of being decisive. The term is borrowed from mechanics, indicating a point against which a level of effort, such as a push, will accomplish more than that same level of effort could accomplish if applied elsewhere. Clausewitz called it the "hub of all power and movement."[3]

Every level of warfare has a center, or centers, of gravity. If several centers of gravity are involved, force must be applied to all if the object is to be moved. Perhaps the most important responsibility of a commander is to identify correctly and strike appropriately enemy centers of gravity. In some cases, the commander must identify specific reachable centers of gravity, if he has neither the resources nor the authorization to act against the ultimate centers. In any event, theater operations must be planned, coordinated, and executed with the idea of defeating the enemy by striking decisive blows.

The theater commander normally will have at his disposal air, sea, and land forces. After identifying the enemy center of gravity, the theater commander must decide which, or which combination, of available forces to use. If he decides to use more than one, he must assign missions to each participant. In the process, he must keep an open mind. He must avoid making an automatic decision that all his available services must participate equally (or conceivably at all), that one is a priori supreme and must be supported by the others, that all must be about the same business at the same time, or that an enemy action demands a reaction in kind.

Instead, he must realize that the nature and objective of the war, and the nature of the enemy will suggest the forces needed for success. On some occasions, one arm will suffice, while at other times all three must be used in any of a wide combination of ways. Sometimes, a particular arm may be the only one capable of carrying out a mission normally associated with another. We will deal with this concept in detail in chapter 9, but let us introduce the idea now by considering two examples, separated by two millennia, which illustrate the theoretical viability of the idea.

When Alexander the Great embarked on his campaign against Persia, the success of his operation depended on securing control of the Mediterranean Sea. Normally, one would assign naval forces to this chore, but Alexander's fleet was too weak to overcome the Persian fleet and had no prospect of becoming significantly stronger. Alexander noted that the center of gravity of the Persian fleet was its shore bases. The center identified, his campaign plan was obvious—before plunging into Persia, he would use his army to seize Persian bases around the Mediterranean littoral. Alexander executed the plan and destroyed Persian sea power without ever winning a battle at sea.[4]

More or less the opposite occurred in the sixteenth and seventeenth centuries, when Spain and England vied for world dominance. The Spanish opted for a land invasion of England that depended on avoiding or overcoming English sea power. The defeat of the Spanish Armada, when it tried to cross the English Channel, wrecked hopes for an invasion. On the English side, the options were either to invade Spain or destroy her economy by using sea power to isolate her from the source of her wealth in the Americas. England opted for the latter and succeeded.[5]

For millenia, then, single arms have at times prevailed, and response—or attack—in kind has not always been the right thing to do.

If in a particular theater, higher levels of command have directed defeat of the enemy through destruction (or disarming through maneuver) of his forces—or if direct action against enemy fielded military forces is actually the best course—the question arises as to which forces must be destroyed and in what order.[6] If equipment, doctrine, or will suggest that the enemy will never use, or effectively use, his air forces, then it would be pointless to expend great effort to destroy them merely because of one's own doctrine. In this case, the air arm could immediately find use in some form of direct attack, interdiction, or close air support.

Conversely, if the enemy believed that either his air force was key to success or that his ability to provide a specified degree of protection against air attack was a prerequisite to continuing the war, then the prime objective might well be the attainment of air superiority. As we will see, Japan surrendered after she lost her ability to defend herself against American air power, and the North Vietnamese accepted a truce under similar circumstances.

In the next several chapters, we will examine air superiority in great detail. We will see the benefits that flow from attaining it—and the penalties exacted for losing it. These chapters, in turn, will lay the base for planning and executing a successful air campaign.

1

Air Superiority—
The Concept

Air superiority is a necessity. Since the German attack on Poland in 1939, no country has won a war in the face of enemy air superiority, no major offensive has succeeded against an opponent who controlled the air, and no defense has sustained itself against an enemy who had air superiority. Conversely, no state has lost a war while it maintained air superiority, and attainment of air superiority consistently has been a prelude to military victory. It is vital that national and theater commanders, their air component commanders, and their surface component commanders be aware of these historical facts, and plan accordingly.[1]

To be superior in the air, to have air superiority, means having sufficient control of the air to make air attacks—manned or unmanned—on the enemy without serious opposition and, on the other hand, to be free from the danger of serious enemy air incursions. Within the category of air superiority, there are variations.

AIR SUPREMACY AND OPERATIONS

Air supremacy, for example, means the ability to operate air forces anywhere without opposition. Local air superiority gives basic air freedom of movement over a limited area for a finite period of time. Theater air superiority, or supremacy, means that friendly air can operate any place within the entire combat

theater. Air neutrality suggests that neither side has won sufficient control of the air to operate without great danger.

The contention that air superiority is a necessity to ensure
victory or avoid defeat is based on theory and on an analysis of
the last half century of warfare. Theory alone would suggest that
it is just not possible to succeed with surface warfare if the surface forces and their support are under constant attack by enemy
air vehicles. And, indeed, the theory is supported by copious
historical examples, a few of which should suffice to make the
point.

Germany destroyed Poland's air force in the first days of the
campaign.[2] From then on, the Poles came under constant and
merciless attack from the air whether they were in bivouac, on
the march, or engaged. Because the Germans had air superiority, they were able to disrupt Polish schemes of maneuver, impose heavy casualties directly on their enemy, and simultaneously facilitate the movement of Germany forces on the ground.
Nine months later, Germany did the same thing in France; the
Luftwaffe won air superiority in two days.[3]

A year later, the German attack on Russia was a classic example of seizing air superiority with massive, violent attacks.
The Germans capitalized on their air superiority by moving
ground forces unprecedented distances up to the late fall, when
weather and failure to follow up on the initial air victories helped
bring the great offensive to a halt.[4]

The attack on Russia had followed, and was a function of,
Germany's failure to win the Battle of Britain and thereby establish the air superiority which was a prerequisite for invasion.[5]
The invasion of Russia was the last instance when Germany was
able to establish air superiority over an opponent. It was the last
strategic offensive Germany was to make before her own homeland lay devastated and occupied.

On the other side in World War II, the Western Allies
achieved air superiority before German Field Marshal Erwin
Rommel's last offensive at Alam Halfa. Rommel observed that
"anyone who has to fight, even with the most modern weapons,
against an enemy in complete control of the air, fights like a
savage against a modern European army."[6]

Rommel subsequently made a similar comment about the situation in Sicily and in Italy. "Strength on the ground was not unfavorable to us," Rommel said. "It's simply that their superiority in the air and in ammunition is overwhelming, the same as it was in Africa."[7]

The value of air superiority was even clearer in the Normandy invasion. Von Rundstedt, the German commander in France during the invasion, reported, "The Allied Air Force paralyzed all movement by day, and made it very difficult even by night."[8]

In the summer of 1944, the Allies gained control over the skies above Germany. By the end of the war, the situation was so bad, because of the incessant bombing permitted by having control of the air, that the Germans had no fuel for their airplanes and only enough gas to give a tank enough for it to make one attack.[9]

Lest it be argued that World War II is ancient history and thus no longer applicable, consider a few cases from wars since then.

In Korea, Lieutenant General Nam Il, the chief representative of the North Koreans at the armistice talks, remarked in a moment of candor,

It is owing to your strategic air effort of indiscriminate bombing of our area, rather than to your tactical air effort of direct support to the front lines, that your ground forces are able to maintain barely and temporarily their present position.[10]

The "indiscriminate bombing" to which General Nam Il referred was a direct consequence of air superiority all the way to the Yalu River.

The Israelis have well illustrated the power of air superiority. In 1967, the Israelis destroyed the Egyptian and Syrian air forces on 5 June and then proceeded to lay waste the Egyptian army in the Sinai, where Israeli command of the air had made life intolerable for the Egyptian soldier.[11]

Six years later, the victors of 1967 paid a terrible price for not gaining air superiority in the first phase of the war. Only after recognizing the need to suppress enemy missile systems—

their primary barrier to air superiority—were they able to turn the tide of battle and go on to win the war.[12]

Finally, the North Vietnamese were unable to conduct a successful conventional offensive as long as American air power was stationed in Indochina. Only after the Americans had left was the North able to mount a decisive ground offense into South Vietnam. In this case, South Vietnamese air attempted little and was easily repulsed by North Vietnamese mobile ground-based air defense systems.[13] As air played no significant role in the invasion for either side, the ensuing action was essentially as it would have been before the era of the aircraft.

AIR SUPERIORITY CRUCIAL TO SUCCESS

In affairs such as war that are only roughly subject to scientific analysis, and where so much depends on the human element, a hypothesis is virtually impossible to prove. However, if one argues that air superiority is crucial to success (as the weight of historical evidence overwhelmingly suggests), then explaining how the operational commander goes about achieving it becomes necessary.

If air superiority is accepted as the first goal, then clearly all operations must be subordinated—to the extent required—to its attainment. This observation is not meant to suggest that no operation be undertaken until air superiority is won. It does, however, mean that no other operation should be commenced if it is going to jeopardize the primary mission, or is going to use forces that should be used to attain air superiority. As with most things, exceptions abound, although when it seems most obvious that the rule should be disobeyed, it is most likely that it should not be.

One may be in such dire straits, brought about perhaps by a surprise attack, that no choice is available but to throw everything into the breach in a desperate gamble to buy some time, or to save some strategically important entity.

The Israelis were faced with this kind of problem in 1973, when they were surprised by both the Syrian and Egyptian attacks. The Egyptian attack was not immediately threatening, but the Israelis judged the Syrian attack as very dangerous. The Israeli high command committed aircraft against the Syrian ground forces, even though the enemy had de facto defensive air superiority over his own lines by virtue of his surface-to-air missile systems. As desperate as the gound situation was, the Israelis quickly realized that they could not continue to use their air force against the Syrian tanks in the absence of air superiority. Consequently, they made the missile fields the primary target, won back air superiority, and then brought the full brunt of their air force against all elements of the Syrian offensive.[14] We will examine further the theory of the emergency situation in chapter 10, on planning an air campaign.

While exceptions may exist, they should not be made the basis of planning. In normal circumstances, air superiority is the first and most compelling task. One normally thinks of attaining air superiority through a combination of aircraft and surface-to-air missiles or guns. Indeed, these two elements normally will play a key role—but by no means the only role. Army ground forces and naval surface forces can and have made major contributions to the air superiority mission. Their contribution can be even greater if they are consciously integrated into the air superiority campaign. This subject will receive expanded treatment in chapter 9, but for now a few examples will help elucidate the idea.

Hitler, in his Directive #6 *For the Conduct of the War*, dated 6 October 1939, noted that the *Luftwaffe* could not attack England from Germany because of range and fuel costs. On the other hand, if Germany occupied the Low Countries, "in no doubt, Great Britain could be struck a mortal blow [by the *Luftwaffe*]." He further saw destruction of the British and French ground forces as "the main objective, the attainment of which will offer suitable conditions for the later and successful employment of the *Luftwaffe* [against Great Britain]." Thus, the seizure of territory to support (and deny) air bases became a ground objective and influenced the planning that went into the attack on France.[15]

On a much smaller scale, the British launched a commando

raid on a small German bomber unit on the island of Crete that had destroyed an inordinate amount of shipping.[16]

Naval forces have reversed traditional roles on more than one occasion. In the 1973 Arab-Israeli war, Israeli gunboats attacked Egyptian surface-to-air missile systems on the Egyptian left flank to pave the way for Israeli air force movements through the opened corridor.[17] Thinking that air superiority must be obtained by air means alone seriously limits commanders in their quest for victory.

Attaining air superiority is not simple in either concept or execution. To begin the process, one must know that there can be a variety of circumstances under which the air battle is joined, and one must understand one's own position before engaging. Otherwise, it is possible to fight a battle well planned for the wrong circumstances. And fighting in the wrong way at the wrong time could well be disastrous.

Three basic factors can affect an air superiority campaign: materiel, personnel, and position.

• *Materiel* encompasses aircraft, surface-to-air weapons, manufacturing facilities for both, and supplies necessary to sustain them. It also includes the infrastructure necessary for their direct support.

• *Personnel* primarily means the very highly skilled people who man combat systems, who have special talents to begin with, and who require extensive training before becoming useful in battle. Pilots and other aircrew members are the most obvious component of this category.

• *Position* summarizes the relative location and vulnerability of air bases, missile fields, ground battle lines, and infrastructure.

All these factors taken together determine the framework of the battle and the options available to fight it.

The three factors can combine in such an infinite variety of permutations as to make analysis futile—unless they are deliberately simplified and put in terms understandable by the commander or staff who must do something with them.

THE FIVE CASES OF WAR

To simplify analysis of the air situation, and to establish a framework for planning, we can divide most wars into one of five cases that are defined by the relationship between the opposing air forces.

In the first case, *Case I*, both sides have the capability and will to strike at each other's bases. This case was the situation in the Pacific in the first part of World War II, when both Japanese and Allied forces could and did strike bases behind each other's lines.

The second case, *Case II*, occurs when one side is able to strike its enemy anyplace, while the enemy can do little more than reach the front. Case II is typified by the Grand Alliance of the United States and Great Britain against Germany after 1943. From that point on, the Allied air forces were able to attack Germany without fear of militarily significant ripostes by the Germans. Case II also suggests that war involves phases. A war that starts out with a particular air situation may not end with the same situation prevailing. Phasing will be discussed in subsequent chapters.

Case III is the reverse of Case II and is a dangerous situation. Here, one side is vulnerable to attack but is unable to reach the enemy. It is the situation in which Britain found herself during the Battle of Britain. She did not feel she had the capability to strike the *Luftwaffe* fields in France; thus, for practical purposes, German bases were safe during the two months of the battle.[18]

The fourth case, *Case IV*, describes the situation in which neither side can operate against the rear areas and air bases of the enemy, and in which air action therefore is confined to the front. Case IV is best illustrated by the Korean War, where the United States imposed on itself political constraints which prohibited operations against Chinese fields and infrastructure north of the Yalu River. The Communists, on the other hand, were unable to attack American fields effectively.

The last case, *Case V*, could come about through mutually agreed political constraints or because neither side had any air

Table 1
AIR SUPERIORITY CASES

Case	Blue Air Fields and Rear Areas*	Battle Lines**	Red Air Fields and Rear Areas
I	Vulnerable	Reachable	Vulnerable
II	Safe***	Reachable****	Vulnerable
III	Vulnerable	Reachable	Safe
IV	Safe	Reachable	Safe
V	Safe	Unreachable	Safe

*Blue and Red fields encompass supporting infrastructure such as power, fuel, and command and control facilities.

**Normally the ground front, but could be a border.

***Safe means that fields are not likely to be hit either because the enemy is unable to hit them, or chooses not to do so, or they are protected by political constraints.

****When *Case II* progresses to its logical conclusion, Red will probably be unable to reach even the battle lines.

power. For example, proxies of two great powers might meet in a place where neither power chose to provide combat aircraft. Clearly, either side could change the rules; thus, it would be useful for participants to anticipate that possibility. Similarly, a war between two poor countries might not involve any significant air activity. Again, though, commanders on both sides would be prudent to think about what would happen if air forces did arrive.

Table 1 summarizes the five cases just discussed. Subsequent chapters will deal with each one in detail.

The five cases discussed here provide an overview of the situation prevailing at the start of a campaign or phase. The commander or planner needs such an overview. However, within its context, the commander or planner must realize that variations in numbers of personnel and materiel support will affect planning significantly. Table 2 provides a simple matrix of some of the possible relationships between materiel and skilled personnel.

Air superiority variables will be addressed throughout this

Table 2
AIR SUPERIORITY VARIABLES

	Skilled Personnel*	Materiel**
A	Limited***	Limited
B	Limited	Unlimited
C	Unlimited	Limited
D	Unlimited****	Unlimited****

*Skilled personnel include those whose training is long and arduous and who cannot be replaced quickly when lost. (Pilots, other aircrew, and technicians.)
**Materiel includes aircraft, missiles, manufacturing facilities, and supporting infrastructure.
***Limited and unlimited are relative to the combatants.
****Must be evaluated over time. That is, both personnel and materiel may be in short supply at the start of hostilities, but may become unlimited either through mobilization, inter-theater transfers, or outside assistance.

book. But like the air superiority cases, a brief review of historical examples should help to make the importance of these variables clear.

Illustrating the situation where both sides have had limited personnel and materiel are the Arab-Israeli wars, where the presence or absence of outside supply has affected the strategy of both sides, and in some ways has accentuated the importance of mutual limitations.

The British during the Battle of Britain offer a good example of the second situation. British aircraft production rates outstripped German production by a wide margin and also comfortably exceeded loss rates.[19] However, the Royal Air Force was below establishment in pilots at the start of the conflict, and the training of new pilots failed to keep up with losses at the height of the battle.[20] The situation might have been untenable had not the battle taken place over Britain, where pilots who bailed out of stricken fighters frequently were able to fly again—in some cases even on the same day.

The United States in the 1980s typifies the third situation, of unlimited pilots and limited aircraft. Whereas the United States has huge reservoirs of pilots who saw service in Vietnam and who could be retrained quickly, it has a very fixed number of aircraft and no way to make fast, militarily significant increases in production.

The situation in which one side has comparatively unlimited materiel is illustrated by the Russian position in World War II—although the Germans certainly didn't believe it or know it until they had been at war with the Russians for two years. The Russians lost nearly 2,000 aircraft on the first day of the war—nearly a third of their total air force and about the same number as the Germans had on the entire eastern front.[21] The Russian loss rate continued on an unprecedented scale until bad weather arrived in October. By mid-1944, however, the Russians had a 6-to-1 advantage over the Germans and seemed to have no problem manning their large armada.[22]

Attaining air superiority means eliminating by one means or another enemy forces that can interfere with air operations. As previously noted, air, sea, or land forces can be used to attain air superiority. In very general terms, two categories of systems can interfere with air operations—that is, block the attainment of air superiority. These systems are *aircraft* and *ground-based weapons*. In support of these weapon systems are detecting systems (such as radar) and electronic countermeasure systems that interfere with or fool opposing electronic systems. These systems are directly related to combat.

Not directly related to combat, but nevertheless essential to it, is the infrastructure that supports these combat systems. The infrastructure ranges from bullets and fuel for the aircraft, to petroleum refineries and the laboratories where scientists work out countermeasures against the newest electronic threats. Depending on the situation, winning air superiority may be possible by eliminating one small part of the enemy infrastructure. In other cases, launching an all-out assault on virtually every part may be necessary.

Regardless of what may be needed to attain air superiority, various ways of going about it are available. For example, one might conclude that elimination of enemy aircraft is the key, but this conclusion does not necessarily mean that enemy aircraft should be targeted directly. The enemy may rarely fly across his own lines, and his side of the lines may be protected by a missile screen. To fly rashly at the enemy's airfields and aircraft without first destroying, suppressing, or circumventing the missile defenses might turn out to be costly at best, and catastrophic at worst.

Simply, in war the shortest distance to a goal may not be a direct line.

The central point of this chapter has been the overwhelming importance of air superiority. For the last half century, air superiority inevitably has spelled the difference between victory and defeat. Commanders and their staffs must consider air superiority in their planning and execution. The framework for analysis suggested in this chapter should make it easier to conceptualize the problem and develop an appropriate scheme for achieving dominance in the air.

Offense or Defense—
the Chess Game

Air superiority, even when not an end in itself, accomplishes two things: It permits offensive air operations against any enemy target at a reasonable cost, and it denies that same opportunity to the enemy. We will start our examination of how to win air superiority with the *Case I* situation in which both sides are equally vulnerable at the start of the war or phase of operations.

Whichever side first wins air superiority will reap significant and perhaps overwhelming advantages.

EMPHASIZE DEFENSE, OR CONCENTRATE ON OFFENSE

In very broad terms, two theoretical approaches to winning air superiority exist, starting from a mutually vulnerable position. The first is to put the emphasis on defending against enemy air, and the second is to concentrate on offensive operations that will reduce the enemy's air capability directly and force him to devote more of his resources to defense.

Naturally, some combination of these two extremes can be available; unfortunately, when they are combined, the availability of forces and time for both necessarily decrease. In fairly close encounters, as Case I wars are likely to be, any decrease in effort, any failure to concentrate, may be quite dangerous.

The first theoretical possibility is defense, but defense has associated with it many problems difficult to overcome. First, it normally requires more than one aircraft to destroy one enemy in aerial battle.[1] Second, from an air commander's standpoint, defense tends to pass the initiative to the enemy. Relinquishing the initiative tends to make defensive concentration difficult, unless bases are positioned for mutual support and the warning of impending enemy attack is sufficiently long to allow massing of defensive fighters. Finally, aircraft awaiting enemy attack are not accomplishing anything—they are putting no pressure on the enemy.

Despite problems associated with defense, a proposal to deemphasize defense in favor of a strong offense may be seen as risky and difficult to sell to political leaders, who are not trained to understand that the effects of offensive operations might produce good defense faster than purely defensive operations. This problem occurred in World War II, when the Germans began using their night fighters to attack British bombers as the bombers were taking off and assembling for night raids on Germany. The program was showing some results (although not significant, because resources allocated to the mission were too small), but Hitler ordered the program abandoned because the British bombers shot down over England made no impression on the German people.[2]

The most serious drawback to defense, however, is that it is a negative concept—by itself it can lead at best to a draw, never to a positive result.[3]

The second theoretical possibility is an all-out offense to gain air superiority. Here, every aircraft capable of crossing the lines is sent out on missions designed to crush the enemy's offensive capability. (Supression of air and ground-based defenses may be necessary before attacking systems supporting offensive air.)

An offensive approach has many advantages. It keeps the initiative and forces the enemy to react. It carries the war to the enemy. It makes maximum use of aircraft and keeps great pressure on the enemy. Finally, assuming the offensive operations are against an appropriate center of gravity, collateral damage probably will be inflicted on facilities that would be attacked in the next phase of operations.

Whenever possible, the offensive course should be selected—if for no other reason than that it is a positive measure that will lead to positive results.

The power of the offense notwithstanding, a variety of reasons exists why adopting the defense may be sound, despite its inherent limitations. Under some circumstances, a successful defence will lead the enemy to conclude that further offensive operations are too costly. Some chance even exists that he will decide to abandon the whole war effort. Before depending on such an outcome, however, one needs to be very sure that the enemy's military and political will has been correctly read, and that one has the strength needed to take a sufficient toll from the enemy before the enemy does too much damage. Outcomes of this kind have been common in land war, but so far only a few examples exist of the same thing in air war.

The first instance of a successful air defense was the British parry of the German air offensive against Britain in the summer and fall of 1940. The British succeeded in exacting a great enough price from the Germans that the Germans abandoned the air offensive and the planned follow-up cross-Channel invasion.

The second example, and one less clear, is the defense the North Vietnamese put up against the American air offensive. The North Vietnamese managed to hold on long enough to exhaust the political will of the American people, even though the American air force had proved its ability to lay waste the country in the 1972 air offensives. This example demonstrates the necessity to read the enemy will correctly. Had the North Vietnamese misread American will, they would have paid a terrible price: The Americans had the capability to do whatever they chose to do from the air.

One also might adopt the defensive because of some anticipated change in the near future. Perhaps a new ally will sign on if the initial defense is successful. Perhaps equipment in significantly greater quantities will be available to permit a better defense or an offense. Perhaps a defense will allow time to build a reserve for an offensive or counteroffensive operation. Of course, in thinking about this possibility, it is imperative to keep in mind that the key word is "perhaps." If "perhaps" does not materialize, then the situation may be beyond recovery.

In other words, the commander who adopts the defense for these reasons is betting heavily on a future that might not happen as he thinks it will. If no choice exists whatsoever, then the commander must do the best he can. At the same time, however, he must make contingency plans for what he will do if the new ally does not join the cause, if the new aircraft does not arrive, or if the reserve is destroyed by enemy action or by higher military or political authority in his own country.

The latter happened to the *Luftwaffe* on at least two occasions during World War II. Adolph Galland, then General of the German Fighter Forces, received a commitment from Hitler to hold back new production of fighters and training of pilots to mount one mighty counterstroke against American daylight bombers. Hitler, however, reneged twice on the promise—once for a futile riposte against the Normandy invasion, and again, equally futile, in support of the 1944 Ardennes offensive.[4] This subject will be covered later from a different perspective in chapter 8.

PHASING AND DEFENSE

The last reason that may support a defense is phasing. The commander may have reason to believe that he can do enough damage to the enemy through defensive operations to make an offensive more likely to succeed. No examples stand out to illustrate this approach in air war, although it has been done on some occasions, with great success, in land war.

One of the most notable examples in land war was the German decision to go on the defensive in the east in 1914, allow the Russians to penetrate into East Prussia, and then launch a counteroffensive, which culminated in the annihilation of the Russians at Tannenberg. The fact that it has not been done in air war does not mean that it can't be done. In theory, the idea of pulling quantities of the enemy into a position where he can be badly hurt has great appeal. On the negative side is the possibility that the enemy will do more damage than expected with his offensive and thereby make the counteroffensive less likely to work.

While acknowledging the possible utility of the defensive, the operational commander should want to go on the offensive at the earliest opportunity for reasons already stated. He must plan a specific course of action. Some of what he does will be a function of his own strength and that of the enemy and of relative geography. Let us see how General Douglas MacArthur and General George Kenney converted a dangerous situation in the Pacific into a decisive victory by emphasizing the offensive and air superiority.

MacArthur had suffered grievously after the Japanese won air superiority in the Philippines. Conversely, he saw what had happened to the Japanese when they tried three offensive operations without first establishing land-based air superiority.

The first came when the Japanese tried to send a convoy to Port Moresby on the south coast of New Guinea. The convoy was met by American aircraft carriers that did enough damage to the convoy's protecting forces to induce the Japanese commander to withdraw.

In the second case, the Japanese tried to make an overland offensive across the Owen Stanley mountains toward Port Moresby without air support or air superiority. Relatively small Australian ground forces and American air forces were able to stop the offensive and throw it back with heavy losses to the enemy.

The third instance was an attempted Japanese landing at Milne Bay at the east end of New Guinea. Again, American air and Australian ground forces decisively defeated the Japanese, who were essentially unprotected by their own air.

Finally, MacArthur concluded that the Japanese had been able to fight so effectively on Guadalcanal only because the US Navy had been tardy in completing Henderson Field on Guadalcanal. Had the field been completed early in the campaign (as it could have been), aircraft operating from it would have presented the Japanese with a much more difficult problem.[5]

MACARTHUR'S FIGHT FOR AIR SUPERIORITY

Following his early New Guinea experiences, MacArthur gradually came to the conclusion that his operations had to have as their primary goal the attainment of air superiority.[6] This is not to say that MacArthur thought that air superiority in and of itself would win the war—he was convinced that only an army assault on Japan would do that. He did, however, believe that winning air superiority was the key to positioning himself for that assault.

After deciding that air superiority was the objective of his intermediate campaigns, MacArthur, aided by his air component commander, General Kenney (who had played a key role in leading MacArthur to this conclusion), inverted the established order of things, and used his ground forces as an adjunct to air in his quest for air superiority over the Japanese. From 1943 to the eve of the invasion of Japan, and with only one exception, MacArthur used his ground forces primarily to seize bases from which air forces could extend the bomb line. How did Kenney and MacArthur prosecute their air campaign?

American pilots and aircraft had started the war against Japan inferior to the enemy. But by the middle of 1942 they were on at least a par with the Japanese. Therefore, the possibility existed of an aerial war of attrition. Although air battles were important to the final outcome, Kenney worked on the thesis that the best and cheapest place to destroy the enemy was on the ground. He switched from the defensive strategy of his predecessor, Lt. Gen. George H. Brett, to a highly offensive campaign within three days of arrival in theater.[7]

General Kenney's goal was to find and destroy enemy aircraft on the ground. Supporting this main objective were aerial combat missions and attacks on the logistic system that provided fuel, food, medicine, and spare parts to the enemy. The key was the availability of ground forces to seize and hold air bases, from which deeper operations could be conducted. His main princi-

Douglas C-47 transport takes off for the Buna front during the Papua, New Guinea, campaign early in 1943.

ples were concentration and persistence. Kenney believed in mounting the largest possible raids against enemy positions and in attacking persistently until they were reduced to impotence.[8] He also was a master of surprise and deceit. The campaign against Wewak illustrates his genius.

In the early phase of the Allied campaign to take Japanese positions on the Huon Peninsula (Lae, Finschafen, and Salamuau) the Japanese moved a large number of aircraft to their big base at Wewak, some 400 miles west of the Huon Peninsula and out of range of Kenney's fighters.[9] If air superiority were to be maintained and extended, Kenney thought that reduction of Wewak was necessary. He couldn't do it, however, with unescorted bombers. His plan to solve the problem was brilliant. Using special overland troops and paratroops, General Kenney started construction of two fake airfields relatively close to the Japanese positions on the Huon Peninsula. At these fields, he deliberately created clouds of dust so that the Japanese would see the construction activity. They responded appropriately by periodically bombing the fields and apparently preventing occupation by American air units.

Simultaneously, at Tsilli Tsilli, some 50 miles further inland, Kenney started construction of the real airfield. He managed to move fighters into it before the Japanese discovered its existence. He then quickly mounted a mass attack on Wewak with his bombers that could now be escorted by the fighters flying out of Tsilli Tsilli. He took the Japanese by surprise, because they were sure that Wewak was beyond range of American fighters and therefore could not be attacked in strength. In two days of mass raids with nearly 200 aircraft in each attack, he won the decisive air battle of the southwest Pacific by destroying more than 200 Japanese aircraft.[10]

Of even greater importance, Kenney started the process that would shortly break the back of the Japanese Army Air Force. His forces killed so many pilots and technicians that the enemy became unable to mount serious opposition, even though he had plenty of aircraft—but aircraft that could not be flown or maintained.[11]

While operations against Wewak were taking place, the American air forces also were conducting, in conjunction with the Navy, intensive attacks against Japanese shipping. These attacks greatly exacerbated the problems the Japanese had with supply and maintenance.[12]

The Japanese could have done to Kenney what he did to them. Why didn't they? Part of the reason lay with their doctrine—something Kenney exploited to the hilt. The Japanese seemed committed to piecemeal reinforcement. At Wewak, at Rabaul, and at Truk they habitually committed small numbers of arriving aircraft in such a way that they could do little to influence the battle. As Kenney put it, the Japanese "did not know how to handle large masses of aircraft. He made piecemeal attacks and didn't follow them up."[13]

The Japanese reinforced failure, while the Americans concentrated on success. The lesson learned from the futile com-

mitment in small numbers of aircraft newly arrived in the theater may be one of the most important in the war. It will be
addressed in greater detail in chapter 8, which is devoted to
reserves. The Japanese seemed unable to learn from their errors. They were taken by surprise at Wewak because they
thought it was out of range. Less than a year later, they suffered
the same fate for the same reason at Hollandia.[14]

Early in the war, the United States made the mistake of committing forces before they were sufficiently trained or numerous. The Strategic Bombing Survey of the Pacific air campaign
reported that American air commanders frequently "failed to
saturate enemy air defensive capabilities, resulting in a high loss
rate and a bombing effort ineffective both in accuracy and in
weight of effort."[15]

Unlike the Japanese, however, the Americans learned the
lesson and emphasized concentration and mass. The American
air attacks became more successful. Additionally, the large raids
so saturated enemy defenses that American loss rates were typ-
ically quite low. As an example, in the 17 August 1943 attack on
Wewak, in which the Japanese lost 150 aircraft on the ground
alone, the Americans suffered no combat losses. Mass and concentration pay!

Ground-based defenses were not significant in the Pacific
war. They were significant (in the sense that they could not be
ignored), however, in the Arab-Israeli battles of the seventies
and eighties and in the American war against North Vietnam.
They also undoubtedly will exist in great numbers in any future
wars between well equipped opponents. The air commander
must determine how much a threat to offensive operations they
present, whether they must be given priority attention and be
physically attacked, or whether they can be suppressed elec-
tronically while the air offensive continues.

Like many things in war, the answer to these questions may
not be obvious. In 1973, the Israelis reacted on the Syrian and
Sinai fronts much as they had in the 1967 war. That is, the High
Command assumed that they would easily overcome the enemy
antiaircraft missiles and fly with acceptable loss rates directly

against attacking ground forces.[16] How did a country widely ad-
mired for the efficiency of its intelligence service make such a
grievous error?

ERRORS OF JUDGMENT

The key factor that led to this error was the contempt the Israelis
felt for their opponents because of the 1967 and War of Attrition
victories. Misplaced contempt for the enemy is not uncommon:
The Germans greatly underestimated the British and the Russians in World War II, the United States prior to Pearl Harbor
thought the Japanese incompetent, and the US Navy discounted
the air defense capabilities of the Palestine Liberation Organization in the ill-fated retaliatory raids in Lebanon in 1983.

Such errors are easy to make, but they are not forgivable.
How can they be avoided?

The first step in assessing an enemy is a very careful review
of intelligence information, followed by dispassionate war gaming, followed by more intelligence collection and analysis, followed by more war gaming, until the answer is relatively certain
or the time has come to act on best available information. This
cycle—and the necessity for dispassionate consideration of enemy capabilities—applies universally.

If the commander decides that ground-based defenses are so
significant that operations above or around them are impossible
or too costly, he must neutralize them. Neutralization can be
accomplished through destruction of key parts of the system,
through electronic suppression of key parts, through adequate
disruption of the system's command and control, or through isolating the system from its source of supply. Of course, a number
of these approaches can be combined. Operations against an air
defense system are very complex; however, as in other aspects
of war, some general procedures have wide application.

In broad terms, a ground-based air defense system has certain characteristics. It is finite and normally has flanks. It has
some directional orientation, based on expected routes of enemy
attack. It is rarely equally strong throughout its width and
depth, and some areas may be very heavily defended while

other areas are only lightly covered. And finally, it is not mobile in theater terms: Although it may be tactically mobile, to the extent that a battery that was one place yesterday may be a few miles away tomorrow, moving large numbers of systems significant distances in short periods of time to fill gaps blasted in some other part of the line generally is not possible. These characteristics suggest campaigns against the system based on flank attacks, penetration and exploitation, or systematic reduction from front to rear.

The Israelis used a combination of flank and penetration attacks very successfully in the 1973 war. Their missile boats hit the north end of the Egyptian missile line at about the same time that General Sharon crossed the canal and destroyed several batteries by ground attack. The Egyptian line breached and flanked, and the Israelis were able to isolate and destroy individual batteries with relative impunity.[17] By knocking a hole in the middle, and by taking out a flank, the batteries no longer were mutually supporting. They were engaged and defeated in detail.

On the northern front, the Israelis conducted operations designed to stop the flow of missiles to the Golan battle area and force the Syrians to disperse their defenses well behind the lines. To accomplish the former, the Israeli air forces went around the flank of the Syrian lines to strike at missile storage areas and transportation nets. For the latter, they attacked infrastructure targets in Syria. These targets had little military value in a short war, but their enormous political and economic value produced the expected reaction: the Syrians devoted missiles and aircraft to their protection, missiles and aircraft that otherwise could have been used at the front.[18] The indirect approach was again the most effective.

A ground-based defense system must be commanded and controlled. If the command and control centers can be identified and destroyed, the whole system becomes much easier to defeat in detail. Unfortunately, these centers normally are well behind the lines and well protected—although it should not automatically be assumed that their physical protection is significant. The Germans, for example, failed to make concerted attacks against the British sector control stations, because they thought them to be underground. In fact, they were above ground in flimsy buildings.[19] If the command and control centers cannot be reached directly, a worthwhile approach may be to attack their sensors. We will examine command and control in more detail in the next chapter.

In this chapter, we discussed the choice between an offense and a defense. We have seen how two forces could confront each other, with each able to strike the other's base areas. In the Pacific war, we saw one side make the radical decision to fight a whole offensive campaign for air superiority. One side was innovative and determined in concentrating mass on an objective; the other made piecemeal attacks and reinforced piecemeal. The way the Americans won the air war in the Pacific (and the way the Japanese lost it) offer valuable lessons to the air commander responsible for fighting the next war.

Having examined the concepts of defense and offense, when the commander has a choice, we now can explore in detail the pure offense and defense. We will start with the offense.

3

Offensive Operations

A commander must undertake offensive air operations if the battle is to be carried to the enemy, and if political objectives exist beyond hoping the enemy will stop his offensive. It is easiest to envision and discuss offensive operations in their pure state, that is, when every thought can be devoted to the offense without concern for defense.

BEST-CASE SCENARIO

Case II meets these criteria and is the commander's dream. His bases are nearly immune from enemy attack, but he can attack all parts of his enemy's structure. The Anglo-American air offensive against Germany from 1943 until the German surrender in 1945 provides the classic example. The Allied bases in England were practically safe from attack, as the Germans had nothing that could reach them without prohibitive losses.[1] The same situation applied to Allied bases in France after the Normandy invasion.

Case II provides the opportunity for decisive action—action so decisive that the war can theoretically be won from the air.

The most likely reason that one side will have safe bases, while those of the opponent are vulnerable, is a lack of proper equipment in the opponent's inventory. One side may not have aircraft that can reach the other's bases. The enemy may have aircraft with ample range, but may lack the training or equipment required for such attacks. He can find himself in this position either because he never had the proper training or equipment to begin with, or because of losses of forward bases or of aircraft during the course of the war. The latter suggests a point which must be kept in mind: The operational situation can change during the war itself. What was a correct approach on day one may be wrong six days, six weeks, or six months later.

ASSAULTING THE AIR CENTER OF GRAVITY

In Cases III and IV, courses of action are limited, because the enemy can only be met up to and over the battle lines. In Case II, however, the commander may drive directly against the enemy's centers of gravity; thus, the selection of a proper center of gravity against which to direct one's efforts is crucial. The selection process will in part depend on relative strengths. If the commander has overwhelming superiority in numbers, he perhaps can afford to target virtually every part of the enemy's air system, knowing that he will get the job done eventually. With enough numerical superiority, this approach is bound to work (although it may cost far more than necessary and may take an inordinate amount of time).

As the offensive commander's degree of superiority moves to equality, and finally to inferiority, the necessity for an accurate assessment of the enemy's center of gravity becomes more crucial. Indeed, if the offense is inferior in numbers, only one course of action may lead to victory. If the commander makes the wrong choice, he may not have another opportunity to win air superiority. The classic case of making the right decision and the right plan was the Israeli attack on the Arab air forces in 1967. The classic example of choosing the wrong center of gravity was the German attack on Britain in 1940. We will look at both in more detail.

The enemy's air center of gravity may lie in equipment (numbers of planes or missiles); in logistics (the quantity and resilience of supply support); geography (location and number of operational and support facilities); in personnel (numbers and quality of pilots); or in command and control (importance and vulnerability).

Equipment

Each of these points must be further evaluated in terms of its position. In other words, reaching every part of the equipment chain, from manufacture to employment, may not be possible. Refineries may be outside the operational theater, but pipelines and storage tanks within it. A careful analysis of enemy doctrine may highlight significant strengths and weaknesses that can either be exploited or avoided. Let us further consider each of the possible centers of gravity.

The layman tends to associate air superiority with destruction of enemy aircraft. Although valid, it is not the only approach. A potentially vulnerable sequence of events (the aircraft chain) must take place before an aircraft fires a missile or drops a bomb. Raw materiel must be assembled, formed, and moved by some method to a manufacturing plant. At the plant, power from some source enables workers to put together the aircraft itself or some subsystem of it. The aircraft, with all its subsystems, then must be moved to an operational field, where it must be protected from enemy attack while it is being prepared for its mission. Finally, it takes to the air. Theoretically, it is possible to eliminate an air force by successful attacks on any point in this chain.

A short look at this aircraft chain will be instructive—keeping in mind that other similarly interdependent systems, such as fuel or pilot training, also can be attacked.

The most difficult and costly place to attack the aircraft chain is in the air. In the aggregate, one friendly plane can destroy one enemy plane. One pilot in one airplane may well shoot down more than one enemy aircraft in a single mission, but that is rare. The majority of fighter pilots will never down an enemy, although, as technology improves, the chances of one pilot with one aircraft accounting for more than one enemy per mission may increase—assuming that countermeasures don't improve commensurately.[2]

Going back down the chain from the air leads to aircraft on the ground. Under ideal circumstances, the results of airfield attack can be impressive. For example, the Germans destroyed more than 4,000 Russian aircraft on the ground between 22 and 30 June 1941.[3] The Germans had less than 1,400 bombers and fighters on the entire Russian front during this period.[4] The Israelis had similar results from their attacks on Arab air in 1967: With 196 operational combat aircraft, they destroyed almost 400 Arab aircraft on the ground in two days.[5]

The historical experience has been that it is cheaper by far to destroy aircraft on the ground than in the air. Whether circumstances will permit such success, however, is a function of surprise, the state of enemy defenses, and the physical protection given aircraft on the field. Note that the most famous instances of such successes have occurred when one side achieved tactical surprise over the other. In some cases, air superiority may possibly be attained by methodically eliminating enemy air bases, although experience in the major wars of this century indicates that airfields must be attacked persistently and heavily if they are to be destroyed. Light, one-time attacks probably will not eliminate an airfield, but may, for a limited period, keep its aircraft on the ground.

The next step back in the aircraft chain, the movement of aircraft from the factory to operational fields, normally does not present much of an opportunity. Ferry routes generally are on internal lines not subject to attack. Worth nothing, however, is the fact that ferry losses for reasons other than enemy action can be quite high. The Japanese, for example, lost a shocking number of aircraft ferrying from Japan to forward bases.[6]

The next significant step back in the chain is the factory. The production of aircraft may depend on a great many factories that produce engines, ball bearings, airframes, munitions, and fire control systems. Sometimes of even greater importance are the people and facilities that support the factories. Power and transportation are particularly critical: Interviews and studies after World War II indicated that power and transportation were the weakest points in German and Japanese war production.[7]

The last step back is to the raw materiel that goes into aircraft

USAF Photographic Collection, National Air & Space Museum, Smithsonian Institution

German Messerschmitt ME 109s roar over the Channel on a fighter mission over England during the Battle of Britain.

building. The sites of raw materiel production themselves are not normally good targets. Transportation nets to the plants, however, can be very vulnerable, as was the case of Japan in World War II.

That choosing a point at which to attack the aircraft chain is far from easy should be clear by now. Important to keep in mind is that there are many ways to attain an objective, and that the most obvious one— —in this case, attack on aircraft in the air— —is quite likely to be the worst choice. Circumstances will vary with each conflict, but the thing to look for is the place where an investment in attack will yield the greatest return. In some cases, a "panacea"' target actually may exist. Where these can be found, they should be attacked and reattacked with persistence.

One more point needs to be made about the aircraft chain. If enemy production sources are outside the operational theater, as they were for the United States in the Vietnam War, and for the Israelis in their wars against the Arabs, then the problem of preventing additional aircraft or missiles from entering the enemy's inventory becomes either easier or more difficult.

In the Vietnam case, keeping the North Vietnamese from acquiring new equipment theoretically was quite easy, as so much came by sea transport, which had to terminate in a very limited number of ports. Once the United States decided to close the ports, and put enormous pressure on the enemy with the *Linebacker II* attacks, the North Vietnamese quickly ran out of missiles.[9]

In the case of the Israelis, blocking entry of aircraft and missiles into the Arab countries was not feasible; consequently, aircraft and missiles had to be addressed closer to the front—where the cost can be quite high as the Israelis discovered in the 1973 war.

Logistics

Enemy logistics may well constitute the real center of gravity. Aircraft can't fly if they don't have fuel, and they can't accomplish anything if they don't have weapons. Ground-based air defense systems are useless if they have no missiles to fire, and neither ground nor air systems last very long without spare parts. Where is success likely in this area?

If the whole logistics chain is open to attack, the most promising link almost certainly will be petroleum. The whole petroleum cycle, from the initial collection points through the refineries to the end user, is exceptionally vulnerable. In World War II, the Allies did not concentrate on the petroleum chain in Germany until May 1944. By September, Germany's ability to produce aviation fuel had fallen by 98 percent; by December, the German military was in such dire straits from a lack of fuel that it had to depend on the seizure of Allied fuel dumps in order to give the Ardennes offensive any chance of succeeding.[10]

Of course, the Allies attacked every element of German petroleum processing, with special emphasis on refineries and

USAF Photographic Collection, National Air & Space Museum, Smithsonian Institution

B-24 *Liberator* bombers take part in the Ploesti mission from a base in North Africa. The oil refineries of Ploesti, Rumania, were the target of a historic 2,000-mile, round-trip 9th Air Force mission on 1 August 1943.

synthetic fuel plants. If it is not possible to concentrate on the refineries, the vulnerability of the petroleum chain decreases—but it still remains a potentially key target simply because a modern military machine cannot function without fuel.

Fortunately for the attacker, even the movement of refined products in any quantity is difficult. Petroleum products must go by railroad, by road in large tanker trucks, by sea, or by pipeline. All these modes of transport can be struck with great success from the air. Such attacks are most effective, however, when the overall fuel situation is fairly tight. In other words, a particular airfield is not going to suffer from attacks on petroleum production or transportation until its own reserves are low. Thus, user reserves should be knocked out where practical. If not practical, patience is needed, for user reserves normally have enough to last for what may seem like a long time, even after the source is completely destroyed.

Other parts of the logistics base might be attacked, and should be, if careful analysis of vulnerabilities, stockpiles, and substitutes indicates that an attack is worth the cost. For example, a sustained attack on plants producing spare parts or munitions may produce satisfactory results over an extended period of time. If time is important, however, choosing a relatively rugged and probably dispersed part of the logistics base may be an error. Again, regardless of the way in which logistics are attacked, a delay almost certainly will be noted between successful attacks and observable deterioration in the enemy's air efforts. Patience and persistence are key.

The German air attack on Poland in 1939 was the first significant use of air power at the start of a war. The *Luftwaffe* staff had correctly identified the need to attain air superiority in the theater,[11] and assumed that it could be accomplished largely by hitting enemy aircraft on the ground and by attacks against physical facilities of the airfields.

The first part of the assumption proved true, but examination of Polish fields after the Polish surrender produced some unexpected results. Attacks against hangars and runways had little effect and were not worth the effort. Coincidentally, however, German attacks on airfields (and railroads) had so completely destroyed communications that, as German historian Cajus Bekker writes, "There was virtually no effective military command from the start."[12]

Geography

In the Polish case, the Germans succeeded in spite of some misplaced effort. In their subsequent attacks on Britain, they were not so lucky.

The German air campaign against Britain in 1940 has become a classic of how to do things wrong. The Germans set out in the summer of 1940 to win air superiority over Britain. During the course of their two-month campaign, they continually changed their objectives, never identified a real center of grav-

USAF Photographic Collection National Air & Space Museum Smithsonian Institution

British-made *Spitfires* in flight over England show their US Army Air Forces markings. It was not unusual to see British planes with American markings, and American aircraft with British insignia.

ity, and demonstrated a remarkable lack of patience and persistence. Of particular note was the short-lived thrust against Royal Air Force (RAF) bases. Starting in the second week of August, the *Luftwaffe* made RAF bases one of its primary objectives. Part of its effort was wasted, because it was directed against forward operating bases used only for quick refueling and rearming. These bases were relatively easy to repair. Another part of its base attack program, however, was directed against main bases, and it lasted until 6 September 1940.

In retrospect, the airfield attack program clearly was weakening the RAF. The Germans, however, abandoned airfield attack on 7 September 1940 and substituted direct attacks on London, which they thought would force the RAF into the air, to be defeated by *Luftwaffe* fighters. One of the reasons why the airfield attack program was cancelled was because its progress could not be charted—unlike the movement of armies on the ground, where progress is easily depicted. Another reason was an assumption that destroying aircraft in the air was relatively easy and inexpensive.[13] As we have seen, nothing could be farther from the truth.

Finally, and perhaps more compelling than the cold military logic—however erroneous—was an emotional desire to retaliate against the British for their nuisance raid on Berlin. We will cover this error in more detail in subsequent sections.

The inability to measure progress in the same way it is measured for land operations demonstrated a German bias toward thinking in only two dimensions that is common even today. This bias is a heritage of thousands of years of ground wars. On the other hand, the assumption that the place to destroy an opposing air force is in the air demonstrated a typical fighter pilot bias. These biases exist in everyone's mind: The successful commander will be the one who can think with his brain, not with his heart.

The preceding discussion shows the usefulness of airfield attack, but also shows that it requires great patience and persistence. One cannot expect to hit an airfield once and forget about it for the rest of the war. Here is where geography can play an important role in deciding the utility of airfield attack. If enemy fields are isolated and cannot easily provide mutual support, an operation to concentrate against one at a time suggests itself. On the other hand, if the fields are mutually supporting—the enemy can easily concentrate forces against an attack—then an airfield attack program may either be exceedingly difficult, inordinately expensive, or not practical. Normally, of course, airfields will be arrayed in some pattern between perfect mutual support and perfect isolation. In this case, if airfield attack is deemed necessary, planners ought to look for a flank or other weak spot that will permit attack in detail.

Personnel

Equipment, whether aircraft or missiles, is of no value unless skilled personnel are available to use it. The people who operate missile systems or who repair aircraft are highly skilled and cannot be replaced easily. Flight crews are especially precious in war, for their production is so dependent on long and arduous training programs and on such a relatively small pool of eligibles that unexpectedly high losses can lead quickly to reduced capability in the air—even when no shortage of airframes exists.

This shortage of air crews is precisely what happened in Germany in 1944 and 1945. German aircraft production hit record highs in 1944, as a result of a desperate and belated recognition that Germany could not survive Allied command of the skies. Unfortunately for Germany, this remarkable production of aircraft was to no avail, as the pilot training program was overwhelmed in 1944, as was the aviation fuel industry.[14]

Two basic approaches to reducing enemy pilot pools suggest themselves. First, a certain number of pilots will be killed, wounded, or captured in the course of air battles, either deliberately sought or which occur in conjunction with other operations. If a pilot is shot down over enemy territory, he normally is lost, at least for some extended period of time. The exception occurs when an air force is operating without time constraints and can afford to devote a significant portion of its daily sorties to rescue operations. If, however, the flyer parachutes to safety in friendly territory, he may well fly again in a new aircraft that same day—as often happened to RAF pilots during the Battle of Britain.

The second manner in which enemy pilot strength could be reduced is through direct or indirect attack on training facilities. The direct attack faces the same obstacles that plague any airfield attack program. In addition, if geography permits, pilot training bases probably will be located as deeply as possible within a country. An indirect approach (such as an attack on the petroleum system) may be more likely to succeed.

Without question, pilots are key to the ability of an air force to operate in the air. However, in order to get at them, something else may have to be done—such as shooting down enemy aircraft or destroying a key part of the logistics base that supports training. These things affect the air battle more directly, or at least more measurably, than reduction of enemy pilot strength. Therefore, except in cases where the enemy has a very circumscribed pool of pilots, while possessing comparatively large numbers of aircraft, pilot strength probably should not be identified as a center of gravity that can be addressed directly. On the other hand, it is important and useful to keep in mind that more direct attacks may have the secondary effect of reducing pilot strength. These attacks may considerably speed the process of winning air superiority. To the extent that the quantity and quality of the enemy pilot force can be identified, opportunities may exist to accelerate pilot attrition.

Command and Control

Command is the sine qua non of military operations. Without command, a military organization is nothing but a rabble, a chicken with its head cut off. Commands exist at all levels, of course, starting with the company or flight level and ranging up through the wing or division level to the ultimate theater or even national command. Destruction or isolation of any level of command may have a serious—and perhaps fatal—impact on the unit or units subordinate to it. Clearly, command, with its necessarily associated communications and intelligence gathering functions, is an obvious center of gravity, and has been from the earliest times: As the death of the king on the field of battle meant defeat for his forces, so the effective isolation of the command structure in modern war has led to the rapid defeat of dependent forces.

One of the problems facing the commander today, which did not face him even a century ago, is the problem of locating the command structure. In simpler times, the commanding general or monarch was readily identifiable to both friend and foe. His death or capture were immediately obvious to both. The results of his death or capture were likewise immediate. Today, a single commander may well be the key to victory or defeat, if he is especially brilliant or stupid. Both cases, however, are on the margin; in normal circumstances the modern staff is capable of keeping operations going along a fairly broad path. Therefore, it

A B-52 *Stratofortress* maneuvers during a joint exercise, *Bright Star 83*. This long-range strategic bomber, in service since 1957, still forms the backbone of the US Air Force's Strategic Air Command.

is not so much the commander who is the center of gravity, but the staff system which serves him.

Unfortunately, staffs normally work out of protected facilities that are well behind the front. They are difficult to destroy or capture. Additionally, within reasonable parameters, individuals on the staff can be replaced by other officers serving elsewhere. Since physical destruction of the staff is difficult and not necessarily long-lasting, other approaches to vitiating command must be considered.

To function effectively, the commander and his staff must receive reliable information on what is happening on both sides of the front, and they must have some way to pass direction to subordinate—and superior—formations. In between receiving and sending, the command element must make decisions. Consequently, command can be attacked in the following three spheres: the information sphere; the decision sphere; and the communications sphere. If any one of these can be sufficiently disturbed, the effectiveness of operations will begin to decrease dramatically. How much it will decrease will be a function of the situation and the pressure being exerted by the enemy.

If the enemy is attacking or defending in a slow, conventional, methodical manner, or if one's own forces are behaving similarly, the urgency of receiving information, making decisions, and transmitting directions is considerably reduced. In fact, absent stress, lower echelons of command need little guidance from higher echelons and probably could continue to function for some time without any guidance. In these circumstances, attacks on the command structure are not likely to produce immediately dramatic results, although something significant may happen over time if a high level of command is knocked out. Conversely, if the enemy is attacking or defending imaginatively with a high tempo of operations, or if one's own forces are doing so, the need for information, decision, and communications goes up exponentially. Now, even a slight disturbance in the command process can be dangerous or even catastrophic.

Command is a true center of gravity and worth attack in any circumstance in which it can be reached. It is vital to remember, however, that results may not be evident for some time unless the enemy is under severe pressure. The destruction of a command element may not immediately be evident on the battlefield, simply because inertia, if nothing else, will allow subordinate units to continue some operations. Patience and persistence again are imperative.

As an illustration, the Germans in the Battle of Britain decided that radar sites were key targets. The Germans launched coordinated attacks on British radar stations early in August and succeeded in destroying one. The British, however, sent false signals from the location of the destroyed radar station to make the Germans think their efforts had been for naught. The Germans responded precisely as the British hoped and dropped attacks on radars on express orders from Field Marshal Goering, the political and military chief of the *Luftwaffe*. The Germans, due to their impatience and lack of persistence, and British deception, stopped attacks against the very thing that allowed the British to mount an effective defense.[15]

The three elements of command—information gathering, decision, and communication—can be attacked individually or together as part of the effort to win air superiority. Each of these elements can be attacked directly or indirectly: The best course will depend on the situation. The decision element is clearly the key, for without it the other two are worthless. Unfortunately, the decision element is the most difficult to reach directly. Normally, the other two elements will offer the best possibilities. Some examples will illustrate problems and opportunities.

The German attack on British radars in the Battle of Britain was an attempt to take out British information gathering. As noted, the attempt failed because of the British deception operation, which in itself was an attack on German command.

By feeding bad information back to the Germans, the British induced the Germans to make a bad decision. Later in the war, the British and Americans attacked German radars indirectly by dropping chaff to mask the direction and extent of bombing raids over Germany. Every major conflict since World War II has seen the use of chaff as a means of depriving the enemy of information.

Attacks on the decision element of command are limited only by the imagination. They can range from direct strikes at enemy command posts to complex operations to mislead the enemy and induce him to do something inappropriate. The former are self explanatory. The latter benefit from illustration.

In the Battle of Britain, the Germans began by more or less concentrating against various parts of the British fighter force. They were making progress when the British made a bomber attack on Berlin. The attack was militarily insignificant, but it was a prime factor in inducing Hitler to direct attacks against London. The shift to London took much of the pressure off the British air force and allowed it to concentrate all of its efforts against the *Luftwaffe*. British bases no longer needed fighter protection, and the Royal Air Force was able to concentrate its fighters against the now predictable *Luftwaffe*.[16] In this instance, no evidence suggested that the British attacked Berlin so that the Germans would attack London. However, it certainly suggests the possibility of doing something to the enemy that will induce him to react illogically.

Another similar case is the Doolittle raid on Tokyo in 1942.[17] The raid itself was militarily inconsequential, but it led the Japanese to the unwarranted conclusion that their home defenses and their defensive perimeter needed to be expanded and strengthened.[18] As a result, the planned invasion of Midway quickly gained great support, the army opened a campaign in east China to capture the airfields where the Doolittle bombers had intended to land, and a total of four fighter groups badly needed elsewhere were held in Japan for home defense until the end of 1943.

All three of these actions had subsequent repercussions—although none so great and dramatic as the defeat at Midway, which was a turning point in the Pacific war.[19] The Japanese reaction to 16 bombers is another indication of how irrational decisionmaking can be, and how vulnerable it is to manipulation.

To date, no really good examples exist of successful theater attacks on just the communications part of the command system. However, some examples are available from both ground and air war at the tactical level that have wider application. Following the heavy air attacks in conjunction with the Normandy invasion, Field Marshal von Rundstedt, the German commander in the west, relayed that "high losses in wireless equipment by fighter bomber attack . . . were noticeable in making reporting difficult."[20]

A more modern example, which combines all three elements of command, is the Israeli operation in Lebanon in 1982. In the late spring of 1982, Israel decided to eliminate the surface-to-air missile system the Syrians had installed in the Bekaa Valley. The Israelis used a variety of innovative ideas and equipment, including army artillery to take out close-in radars, drones with television cameras to give the commander a real time view of the battle, and F-15 fighters in an airborne control role. The sequence of the operation was roughly as follows: The Israelis

USAF Photographic Collection, National Air & Space Museum, Smithsonian Institution

A B-25 *Mitchell* medium bomber takes off from the deck of the US Navy Aircraft Carrier USS *HORNET* with heavy seas and overcast sky on the last leg of the Doolittle raid on military targets in Japan on 10 April 1942.

fed bad information to the Syrians by launching remotely piloted drones that produced radar returns similar to fighters. The Syrians shot SA-6 missiles at nothing and exposed themselves to various types of fire. Next, the Israelis reduced Syrian information gathering capability significantly by hitting Syrian radars, to open the way for attacks on individual missile sites.

After the missiles failed, the Syrians launched waves of fighters to intercept the Israelis, but the Israelis jammed the data and voice communications on which the Syrian pilots were dependent.[21] As a result, the Syrian fighters were reduced to uncontrolled singles trying to operate against superior numbers of well-controlled enemy aircraft.[22] The loss of information-gathering systems and communications led the Syrian command to throw more fighters fruitlessly into the fray. In the week-long operation, the Syrians lost 85 fighter aircraft and 29 surface-to-air missile batteries.[23] The Israelis lost nothing to Syrian air and, depending on sources, two to three aircraft to ground fire.

The Israeli success was phenomenal. It was the result of effective attack on all elements of Syrian command. Whether that success can be duplicated on a much broader front is a matter of speculation; however, a broader operation even 20 percent as effective would still be spectacular. The opportunity is there and when, as in the case currently under discussion, one's own bases are relatively secure from attack, possibilities are so enormous that the utmost attention should be given to a concerted attack on the enemy's command system.

AIR DOCTRINE

In deciding where to put the emphasis of the air superiority campaign, making a careful analysis of enemy doctrine is important. In the process, avoiding ethnocentricity is especially important; one must not assume that what one's own Service considers logical and necessary is what the enemy will consider logical and necessary. In a certain sense, war through the ages has been a battle of doctrines. The really decisive successes have come to those who adopted a new doctrinal concept to which their enemies were unable to respond: The refused center at Cannae baffled the doctrine of line clashes; the longbow at Crecy[24] beat the doctrine of the heavy cavalry charge; the tank, to some extent in World War I and markedly in World War II, defeated the doctrine of linear warfare; and the doctrine of air bombardment brought crushing defeat to the countries whose doctrine depended on armies and fleets for protection or conquest.

Examples abound from World War II to the present of air doctrine, for good or ill, playing a major role in the outcome of battles and wars. The classic case is that of the Germans, who although judged apparent masters of air doctrine after the great successes in Poland, had such fundamental failings that they were virtually precluded from winning the wars on which they had embarked.

German

Development of air doctrine began in Germany in the 1930s. The *Luftwaffe*, like other major air services, was intrigued by aerial bombardment ideas espoused by Giulio Douhet,[25] in his epic book *Command of the Air* (1921; English translation, 1942). Hardware development was moving apace, with a four-engine bomber on the drawing boards that was to have sufficient range to penetrate beyond the Urals.

In 1936, however, the man who provided the broad strategic vision for the *Luftwaffe*, its Chief of Staff, General Walther Wever, died in a plane crash. The officers who followed him did not have his broad vision and administrative ability. In addition, the head of the Office of Air Armament after the Spanish Civil War believed that all bombers had to dive in order to have satisfactory accuracy. Compounding the issue were technical problems in bomber engine development.[26]

These factors meant that Germany had a short-range, tactical air force when she went to war against two countries, Britain and then Russia, which could only be reached with long-range aircraft.

Lack of a long-range air force meant that Germany had to meet and defeat her enemies at the front—the same way that armies had opposed each other for centuries. Such a course also has worked for centuries, but against enemies similarly constrained or against enemies with forces not too numerically superior. In World War II, these conditions did not prevail for Germany. On her western front, the British and Americans had long-range air forces that could attack every facet of German life, from the factory to the front. All of Germany was under bombardment by 1944, whereas British and American rear areas were practically immune to *Luftwaffe* bombing. On the eastern front, Germany faced the Russians, who had moved their industry behind the Urals and greatly out-produced Germany in military equipment. The Germans could not reach Russian factories, so were compelled to take on Russian equipment at the front where numerical superiority eventually proved overwhelming. In essence, Germany entered a war with doctrine and equipment that were inadequate to the task.

Contrast Germany's mismatch of strategy, doctrine, and equipment with American experience against Japan.

American Air Doctrine

American air theoreticians during the 1920s and 1930s concluded that it was possible to defeat an enemy by direct air attack on his homeland. These theoreticians had put together proposals for the "strategic" bombing attacks on Germany and one of them, General Haywood Hansell, pushed for the same thing in the Pacific. He argued that the seat of Japanese strength was in the home islands; if airpower could so punish those islands by direct attack, then the armies on the perimeter would be little more than useless appendages. To carry out such a plan, however, he had to have bases within about 1,600 miles of Japan, in order to bring the islands within range of the B-29 bombers that would start arriving in 1944. He therefore urged that the previously planned central Pacific thrust be devoted to winning suitable bases in the Marianas. To him, operations in the southwest Pacific were peripheral. After lengthy discussions in Washington, he finally convinced the Air Staff, and then the Joint Chiefs, to approve a campaign with a primary mission of seizing bases in the Marianas from which US forces could conduct intensive air bombardment and establish a sea and air blockade against Japan and from which to invade Japan proper if this should prove necessary.[27] The Joint Chiefs were willing to hedge their bets.

Hansell himself had the opportunity to command the first of the B-29 units based in the Marianas—although he was forbidden to fly missions against Japan and was succeeded by General Curtis LeMay, who directed the fire bombing raids. These operations started in late 1944 and became intense with the fire bomb raids on Tokyo in March 1945. In late spring, the Japanese

USAF Photographic Collection, National Air & Space Museum, Smithsonian Institution

A jeep leads a B-29 *Superfortress* of the 499th Bomb Group, 7th Air Force, to hardstand on Saipan Island, in the Marianas Group, in 1944. The heavy bombers used the islands as launching points for northward strikes to Tokyo.

government started its first tentative efforts to negotiate an end to the war. Shortly thereafter, Japan started looking for ways to surrender contingent only on protection of the Emperor.[28]

Japan surrendered unconditionally immediately after the two atom bombs fell on its cities. This is not the place to debate whether the atom bombs were needed. What is clear is that the Japanese had lost air superiority over the home islands. Previous losses of trained personnel and obsolescent aircraft made it nearly impossible to resist the marauding American bombers. At the time of the surrender, the Japanese had 2 million men and 9,000 aircraft in the home islands alone. Nevertheless, as the Strategic Bombing Survey concluded, "It seems clear that, even without the atomic bombing attacks, air supremacy over Japan could have exerted sufficient pressure to bring about unconditional surrender and obviate the need for invasion."[29]

Unlike the Germans, the Americans had developed doctrine, aircraft, and training appropriate to the problem. They won.

Japan itself was the political center of gravity. If it could be sufficiently threatened, the war would end. The key point here is that any war is almost certain to have a similar center of gravity. If it can be reached directly without reducing the defenses enroute, doing so should be considered.

How to affect that center of gravity is the next question. As the Strategic Bombing Survey suggested, no government can long function when the enemy operates freely above it—that is, when the enemy has air superiority. The same suggestion may well pertain to levels short of national government. The mere possibility, however, means that the air superiority campaign must be given great thought—as an end in itself, or as a means to an end.

Syrian Doctrine

Almost 40 years after Germany and Japan fell in defeat, the Israelis brilliantly capitalized on weaknesses in Syrian doctrine. The Syrian air doctrine was closely modeled on the Soviet doctrine and called for very close control of fighter aircraft by ground stations. This doctrine puts a premium on the ability to gather information and to communicate. This doctrine, in turn, suggests a possible vulnerability. Indeed, if one had only a copy of Syrian doctrine, and no record whatsoever of any Syrian combat experience, one could logically conclude that the command and control system ought to be a particularly lucrative target—as indeed it was in the 1982 battles previously discussed.[30]

As suggested in the beginning of this discussion, the case in which one's own fields and rear areas are safe from attack, while those of the enemy are not, is the best possible situation. Its existence, however, does not ensure victory or make victory come cheaply. Careful consideration of enemy centers of gravity, assisted by analysis of enemy doctrine, is the first step to success. The second step is concentration of effort. Especially in the situation where one seems to have numerical superiority, there is a tendency to try to do everything. In all likelihood, the net result will be that nothing is done as efficiently as it should be.

A very important point to keep in mind is that Case II is different from the other cases. It is significantly different from the case where the enemy's rear areas are not reachable. If a Case II war is fought as though it were a Case IV war (where enemy rear areas cannot be attacked for either military or political reasons), the time to bring the war to a successful conclusion will be extended at best. In other words, if the opportunity to strike the enemy's bases and support systems is available and not taken, the price for this neglect will be high. As logic and historical experience very clearly indicate, the most expensive way to destroy enemy air is to engage it over the front in a head-on battle. One bomb dropped by one aircraft on one factory or power plant may directly or indirectly destroy great numbers of enemy aircraft, whereas one aircraft at the front is unlikely to destroy even one of the enemy. The same principle holds true for interdiction, as will be shown in chapter 6.

Political leaders may be loath to attack enemy rear areas at times. Conceivably, cogent political or strategic reasons may call for avoiding attacks on rear areas. It is imperative, however, that the operational commander make clear to the political authorities that they are directing a *militarily* illogical course, and that the cost and duration of the war almost certainly will be far higher and longer than it otherwise might be.

From the commander's ideal case, we will turn next to the most dangerous of all cases—the case where the air commander is forced to accept the pure defense.

Defensive Operations

For the air commander, Case II, where the enemy's bases are open to attack but one's own are not, is the ideal situation. Conversely, the worst situation (Case III) is where the enemy can operate against one's bases while his are immune. Not only is the Case III air superiority battle the toughest to win, but the consequences of losing it are the most severe, as loss of the entire war becomes quite likely.

The Case III situation can develop in a number of ways. Equipment, such as long-range aircraft, may not be available to carry the war to the enemy. A lack of will may prevent carrying out strikes against the enemy. The lack of will could stem from fear on the part of flyers, or political dreams that restraint may keep the enemy from doing something even worse than what he is already doing. Doctrine may influence or control the situation. Just as theorists in the 1930s were sure that the unescorted bomber would always get through, some think that current air defense systems will suffice and that offensive operations are futile. Even if doctrine provides for offensive operations, quite possibly they have not been practiced in peacetime, and the force consequently is unprepared to take on such a complex and sophisticated operation.

Finally, a variety of circumstances may prohibit an offense. One possibility is that the initial enemy onslaught was so violent that it destroyed the systems or personnel needed to support an attack. In any event, Case III clearly can happen—as indeed it happened to Poland, France, Great Britain, North Korea, the Arab states in 1967 and 1973, and to North Vietnam.

A WEAK POSITION

The defense, in classical land warfare, may well be stronger than the offense, as Clausewitz postulated. In air war, however, the opposite seems to be the case. Several reasons explain this apparent contradiction.

- First, air forces have such tremendous mobility that they can attack from far more directions than can a land army.
- Second, the rapidity with which air forces move makes concentration against them more difficult than concentrating to defend against a land attack.
- Third, the defender on land normally has prepared positions from which he can fire at an attacker who must by definition move across open territory where he is at a decided disadvantage.
- Lastly, when air forces meet in the air, the difference between attacker and defender tends to blur (if not disappear entirely). The lack of difference between attacker and defender in air war has important ramifications for both sides, as we will see shortly.

Historically, being on the pure defense in air matters clearly is fraught with danger. The danger may be greater or lesser, depending on the nature of what has to be defended. Easiest to defend is a reasonably tight complex where defenders can meet challengers any place on the periphery, and where the defenders can provide each other mutual support. Most difficult to defend is a long narrow area where distances preclude mutual support and where the attacker can choose a variety of targets for his thrusts at any particular time. Two points need clarification.

First, we are speaking here of theater-size operations, not about defense of a single airfield, factory, or even city. Second, we are making the assumption that, for the foreseeable future, the only really effective counter to an aircraft is another aircraft.

As we have said before, this argument is not to suggest that ground-based defenses can be ignored or that they are not dangerous. In fact, they are dangerous enough that one must assume that no one will commence an offensive air campaign unless he is relatively sure that he will be able to neutralize them by one means or another.

The relations of mass, or numbers, between the attacker and the defender make geography, or, more specifically, the disposition of airfields, of prime importance for the air defender. For the attacker, mass must be available to do a reasonable amount of damage—again, on a theater basis. True, a single aircraft with a guided weapon can take out a point target, such as a bridge (assuming, of course, that a single aircraft can penetrate the defenses). On the other hand, a single aircraft cannot put an airfield, marshaling yard, or other significant military target out of commission; only a mass of aircraft can do that. Not all air forces have learned this basic principle; thus, some might try to conduct a campaign with small numbers. Should this event happen, the defender can count himself fortunate. It is not wise, though, to plan on the enemy's stupidity. One must expect that any serious enemy will attack with strong forces. Strong forces must be met with stronger forces.

The history of air war, as short as it is, has shown clearly that masses in the air can only be opposed by counter masses. Attempts to defend with inferior numbers (in a particular battle, as opposed to inferiority in the theater), or, conversely, to attack with inferior numbers (on a particular engagement) have been notably unsuccessful.[1] We already have discussed illustrations of this principle from the war in Europe, as well as the war in the Pacific. We will see more.

If mass is important in defense, the problem becomes one of producing mass at the appropriate time. We must adjust our perspective and think in terms of air battles. Mass is only important insofar as it can be brought to bear against an enemy attack. Thus, aircraft that cannot take part in an air battle are irrelevant.

How does one not lose a Case III air superiority campaign? The question is deliberately phrased in the negative, because the fact that one is on the defense means that the best possible outcome is not to lose. Nothing positive can be achieved from defense—although a successful defense may prepare the way for a subsequent offense.

Fortunately, at least one small—very small—advantage exists to being on the defensive. Simply, the enemy's motivation for offense, and thus his willingness to accept punishment, may be less than that of the defender. The attacker is hardly likely to throw his entire air force into the fray and lose it all before deciding to give up the attack. Conversely, the defender might not find it illogical to expend his entire force in an attempt to protect himself. This fact gives the defender a slight psychological edge that can—and must be—exploited.

IMPOSING HIGH ENEMY LOSSES

The key to not losing is to inflict enough damage on the enemy that he becomes unable or unwilling to pay the price. Is this such a truism that it doesn't need stating? While it may be a truism, it is not easily put into action. One necessarily must think exactly what must be done to lead the enemy to give up his offense.

On the defense, the only way to hurt the enemy is to knock down his aircraft and capture or kill his flyers. The number of aircraft knocked down are important, but more important is the timing of their destruction. The enemy certainly will accept some level of losses, and probably has determined that level in advance. One percent is an attrition level that most air forces could sustain without making drastic changes in their campaign plans.

For illustrative purposes, assume an air force of 1,000 aircraft suffers a 1 percent loss each day for 10 days. Total losses would amount to just under 100 planes. If results had been good for that 10-day period, the commander probably would continue operations. But now, let us take the same total loss and inflict it on a single day. Almost every commander, under these circumstances, would seriously reconsider his plans. First, he clearly can't accept losses of that magnitude more than once or twice.

Second, losses of that size almost are certain to have hurt some units so badly that they would have to be withdrawn. Third, his flyers suffer a blow to their morale and to their feeling of invincibility.

In short, the difference between losing a little each day and losing a lot on a particular day is significant. The defense must inflict as many bad days on the offense as possible, even if that amount of action necessitates reduced activity on some days.

In the foregoing example, we suggested that 1 percent was a sustainable attrition rate, whereas 10 percent in a single day was not. The true figures may vary somewhat, but these percentages have historical support.

In World War II, American air forces generally felt that 10 percent was the greatest attrition they could accept without changing something. Indeed, in October 1943, the *Luftwaffe* had its best month of the war and succeeded in imposing a 12–16 percent loss rate on the Americans. That rate was so unacceptable that the commander of the 8th Air Force stopped further deep, clear-weather raids into Germany for almost four months.[2]

Thirty years later, the Israelis lost 40 fighters over the Golan Heights in a single day. This loss rate, more than 10 percent, forced them to stop operations until they could figure out a better way to do the job—despite the fact that the Syrian breakthrough the air force was trying to stop would do great damage if it succeeded.[3]

The goal, then, is to impose very heavy losses on the enemy in the shortest time possible. How can that goal be accomplished? Two general principles must be followed.

• The first is to concentrate forces, to confront the enemy with superior numbers in a particular battle, sector, or time.
• The second is to accept the fact that it is not possible to defend everywhere and everything: He who tries to defend all defends nothing. Penetrations are going to take place. When that fact is accepted, it becomes easier to do the concentrating which will permit significant victories with acceptable defender losses.

CONCENTRATE FORCES

Another phenomenon is important for the air commander to understand: Loss rates vary disproportionately with the ratio of forces involved. Two forces equal in numbers (and reasonably close in equipment and flying capability) will tend to have equal losses when they meet. Keeping the same equipment and personnel, as the force ratios go against one side, that side will have greater loss rates than the changed ratio would suggest. Conversely, for the side for which the force ratios become more favorable, loss rates will fall more than the ratios would indicate. The change in loss rates, either positive or negative, is not linear; it is exponential. Furthermore, no point of diminishing returns for the larger force seems to exist. That is, the larger the force gets, the fewer losses it suffers, and the greater losses it imposes on its opponent.[4]

Unfortunately, no good rule of thumb exists for how much superiority the defender should have over the attacker. A few examples, however, may give some ideas.

- The Japanese attacked Midway with 108 bombers and fighters. Midway's Marine squadron of 26 fighters suffered almost 100 percent losses.[5]
- On 11 January 1944, the American air force attacked a target deep in Germany with a force of 238 bombers and 49 escorting fighters. The Germans opposed it with 207 fighters. Losses were 34 bombers. Just over a month later, on 19 February, a force of 941 bombers escorted by 700 fighters met German opposition of about 250 fighters. In this encounter, the Americans lost just 21 bombers—a lower absolute number and a lower percentage.
- In June 1982, an Israeli defending force of 90 fighters met a Syrian force of 60 fighters. The Israelis had no losses, while the Syrians lost 23 of their aircraft.[6]

Modern weapons might arguably have invalidated the experience of World War II and Korea, and the Israeli battle last cited possibly was an anomaly. While certainly possible, this argument seems unlikely. Lots of aircraft targeting fewer aircraft

5 · 7/10 CLOUD · N · S · E · W. BERLIN AREAS BOMBED ON PFF.

USAF Photographic Collection, National Air & Space Museum, Smithsonian Institution

B-17 *Flying Fortresses* take part in the bombing of Berlin in 1944.

are bound to get better results than the other way around. This conclusion has nothing to do with the quantity versus quality debate. Better airplanes are going to perform better than inferior ones—a fact noted by the great German ace Manfred Von Richthofen in 1918, when he commented, "Besides better quality aircraft they [the British] have quantity. Our fighter pilots, though quite good, are consequently lost."[7]

We said earlier that no fixed ratio exists that the air commander can use as a rule of thumb. We have seen, however, that the greater the ratio of defender to attacker, the more likely is the defense to succeed. The defending commander must ensure that the ratio is in his favor.

All this emphasis on numbers may seem to suggest that the outcome of the Case III air superiority campaign could be judged on the basis of relative prewar strengths—perhaps tempered by production rates after the war started. This emphasis on numbers also might suggest that the defending commander is doomed if he has fewer aircraft than the offense. Neither suggestion is true. Static balances are of interest, but they don't have much to do with how the war is likely to end, unless the numbers are absolutely overwhelming. What counts is the numbers when two forces meet in actual battle.

The smaller defending air force can win if its aircraft are properly employed, and if they are concentrated in such a way as to outnumber the attacker in any given engagement. Concentrating to achieve numerical superiority is imperative, even if doing so leads to some attacks escaping without interception. Far more important and effective than getting a constant 1 or 2 percent a day is imposing heavy losses in one battle or on one day.

Also important is that the defending commander, especially the commander of a force that is overall inferior in numbers to the enemy, recognize that his losses will be lower when he outnumbers the enemy in an engagement. And again, big enemy losses on a single day or on one raid do wonders for morale—on both sides.

Having prescribed concentration and numerical superiority, how is this superiority attained? It will be difficult to attain in war, if it is not practiced in peace. Galland, after concluding that he needed numerical superiority—and desirably 3- or 4-to-1 even in the days before American bombers were escorted—found that his pilots had great trouble operating in formations larger than a flight, because they had not flown them for the three years since the Battle of Britain.[8]

Practice is a necessity. So is creating the mind set of "fight in superior numbers and win." The fighter pilot has a tendency to plunge bravely into any fray, but such action can be wrong. Audacity may lead to defeat. The air force inferior in numbers to its enemy in a theater must fight better and smarter to win. Its generals must concentrate their force in such a way that it has superior numbers when a battle is joined. The political slogan of "fighting outnumbered and winning" has no place at the operational level of war.

IMPLEMENT WARNING AND CONTROL SYSTEMS

The defending air force must devise ways to get many aircraft off the ground quickly. This is especially significant if air bases are not well distributed. None of the foregoing will help, however, if a good, survivable warning and control system is not available. Obviously, the more warning, the better, although the commander must be sensitive to enemy feints designed to draw the defenders into the air and then to strike when they run out of fuel. The enemy is especially likely to resort to subterfuge if the defense has successfully concentrated several times and taken a heavy toll as a consequence.

The strategy and tactics of the enemy will either complicate or simplify the job of concentration. In the Battle of Britain, the *Luftwaffe* initially targeted fighter bases and aircraft production facilities. Since these bases and facilities were scattered over the southeast corner of Great Britain, the British had difficulty knowing exactly where a raid might be headed when it was first detected on radar. The Germans were concentrating their offensive forces. The more they did so, the better their success. For a variety of reasons, the Germans in early September switched their efforts to London, removing doubt as to where raids were headed, and making the job of concentrating the defense that much easier. Also, the end of *Luftwaffe* attacks on British fighter bases made operations from the bases simpler.

Finally, at a time when the Germans thought that their enemy was losing the battle, the British, with the help of the "Ultra"[9] code breakers, were able to concentrate all their forces, including those that normally would not have been committed to the London area, for a mighty attack on the Germans on 15 September. The Germans took such heavy losses, and were so surprised to find that the British seemed stronger than ever (although in actuality they weren't), that they gave up the serious air assault on the island kingdom.

In retrospect, the Germans might have won the air war had they not given the air bases a respite and had they not shifted their efforts to London, where the British could concentrate against them.[10]

One other approach used by the British is worth mentioning, although it will receive more treatment in a later chapter. The British maintained reserves.[11] They frequently rotated units out of the hot spots, allowing them to recuperate. These units were available when the battle reached a decisive phase. While the British were rotating units and thus maintaining a reserve, the Germans were using everything and everybody they had. When the time came for the decisive punch, nothing was left for them to add.

The Case III air superiority fight can be won if the air commander employs his forces well. If he concentrates, if he accepts some penetrations in order to maul others, and if he develops and uses a good warning and control system, he can beat a larger air force. Conversely, if he tries to defend everywhere, if he commits his forces piecemeal, if he fails to concentrate, he will lose—and may even lose against a much smaller air force if the attacker outsmarts him.

Air warfare, especially in the defense, is extraordinarily complex and demanding. Careful thought and cool execution are necessities.

From the pure defense, we move now to those anomalous situations in which air power is not a factor for any of a variety of reasons, or where both sides are forced to fight over the front without attacking each other's rear areas. The commander may be greatly frustrated in both, but the general rules we have discussed so far still apply.

5

Limited Options

So far, we have discussed air superiority cases when the commander was involved in a life or death struggle—either to defeat the enemy or keep the enemy from overwhelming the commander. As explained in chapter 1, however, situations exist in which the air battle is largely confined to the area over the front, or where air power plays no significant role.

Both of these cases require a different perspective.

When the rear areas of both sides are relatively safe, either because of political restraints or because of physical inability to reach appropriate targets, the overall campaign plan is easier to devise—although it may be harder to execute. In this case, air superiority is unlikely to be an end in itself; rather, it is needed to prevent enemy air interference with ground operations over or near the front, while permitting friendly air operations over corresponding parts of enemy territory.

When the enemy rear cannot be reached, options are very limited. To achieve air superiority, little can be done beyond the elimination of enemy aircraft in the air and the suppression of enemy ground-based systems. Under these circumstances, the commander must decide whether the ground-based system constitutes a sufficient threat that it must be attacked first, or whether it can be suppressed by electronic means while enemy aircraft are defeated in the air.

We discussed neutralization of ground-based defenses in chapter 2. Let us now look at the air battle.

OPTIONS DEPEND ON THE ENEMY

If enemy air forces cannot be attacked on their bases, they must be attacked in the air. Options depend on the enemy's strength and doctrine. If the enemy considers himself comparatively weak, he will attempt to avoid aerial combat, while concentrating efforts against aircraft that may be harassing his ground troops or supply lines close to the front. One could even imagine a situation in which great waves of fighters are sent over the lines to engage enemy air, but always return without destroying any enemy aircraft, because the enemy chose not to fight. Should this situation occur, air superiority is won by default and the next phase in the campaign can begin. If the enemy air force is not quiescent, however, it must be met and destroyed by fighter forces.

Fighter Screen

Several general methods are available to run the fighter campaign when the enemy rear cannot be attacked. The first method is a fighter screen between enemy bases and the front. The Americans used this technique successfully in Korea after the Chinese entered the war. The difficulty in such an operation is that it relinquishes the initiative to the enemy. The enemy may or may not choose to challenge the screen, he may hold his attack until he judges the screen is nearly out of fuel, or he may succeed in putting superior mass against the screen. The proper choice depends largely on numbers.

If the enemy is notably inferior in numbers, the air commander may possibly maintain screens on station of sufficient size to cope with any possible enemy attack. Although specific ratios are difficult to prescribe, a minimum of 1-to-1 is the least that prudence demands. Obviously, a permanent fighter screen is a very expensive operation.

If the enemy is equal or superior in numbers, maintaining a fighter screen becomes exceedingly difficult, because the enemy can overwhelm it at a time and place of his choosing. The key here, as always, is to fight superior and win. On anything but a theater basis, talk of fighting outnumbered and winning is dangerous. If a single lesson can be learned in military history, it is that the key to winning battles is to have greater forces at the key location than does the enemy. The trick is to outwit the enemy and thus out-concentrate him at the right time.

One of the major attributes of airpower is its mobility. If that mobility can be used to provide concentration, it can win the battle. In this instance, it may be possible to create a screen as the enemy assembles to attack if bases are close enough to the front, if detection systems are good enough to give sufficient advanced warning of enemy movement, and if large enough forces can be launched quickly from enough fields to give numerical superiority when the battle is joined. To follow this course requires superb detection systems, superb command coordination, and bases close enough, in time or distance, to the battle area. If it can be done, this course can regain some of the initiative as the enemy cannot know whether or how he will be engaged.

ESCORT OPERATIONS

A fighter screen, or the mobile version of it just discussed, may not be needed if the enemy is using his air defensively. Under this circumstance, escort of close air support or interdicting aircraft may be sufficient. As previously noted, if attacks against enemy ground forces are being carried out, enemy air either must answer or de facto relinquish air superiority. Assuming he will accept air battle, the question becomes how to conduct the escort operations. Two basic approaches are available: sweep, and close escort.

In the *sweep* option, the fighters precede the bombers and engage enemy air found enroute or on the flanks. In the *close* escort option, fighters stay very close to the bombers and attempt to drive off the enemy when he attacks. The latter has a long history of failure: the *Luftwaffe* against Britain in 1940;[1] the US Army Air Forces against Germany in 1944;[2] and the US Air Force against the Chinese in Korea and against the North Vietnamese in Indochina.[3] Some future war, however, may reveal that close escort will be the proper approach.

The preceding discussion of sweep and close escort represents an excursion into the world of the tactician, as opposed to that of the operational commander.

The operational commander normally should stay away from tactical problems. Some tactical decisions, however, have enormous impact on the whole war. The sweep versus close escort decision is one. The tactical decision area also is one of those areas where commanders from different branches—even within the same Service—may have radically different ideas about what is proper and what is not. In the escort area, for example, bomber commanders historically have felt unprotected when they could not physically see their escorting fighters. When this sort of rift develops, the operational commander must step in to make a tactical decision.

The distinguishing feature of Case IV is the base area sanctuary enjoyed by both sides. Given this sanctuary, the campaign is likely to turn into a long slugging match, in which either side has difficulty doing anything more than wear the other down. This development is especially evident when both sides have roughly equal numbers and supporting production of weapons and personnel. If one side is notably inferior to the other, in terms of either pilots or aircraft and missiles, that side can only play a very careful game and look for opportunities to do damage to the opponent without suffering large losses to itself. As long as it takes this course, if can stay in the war for a long time. This observation is not to say that its ground forces are not going to suffer horribly in the process—as did those of North Vietnam after the United States entered the war.

The case we have just discussed is one of the easiest to solve from the operational level, because so few options are available. It is apt to be maddening for all concerned, and significant differences may arise with the political leadership if restraints on attacking enemy rear bases are politically motivated or militarily unsound. Should this case happen, the operational commander must give his candid advice as to likely costs with and without the constraints. Although Case IV presents few options to the commander, Case V—in which air power is not significant—presents even fewer options. Even in this case, however, the commander must still think about air power.

A war without combat aircraft is most likely to occur when two relatively primitive or poor forces clash. Less likely, but still possible, might be a phase in a war that takes place after both sides lose the use of their air forces, either because of combat attrition or because of maintenance problems. Regardless of how it comes about, air superiority will not be a problem for either side. We mention this scenario only because the situation could change quickly if one side acquires air power or if a supporting power decides to introduce it.

Situations do change, and the operational commander should run an air planning exercise concurrently with his real ground or sea war planning and execution. The planning should focus on how and where air should be used offensively if it becomes available, and on what targets to defend should the enemy acquire it exclusively. Thinking and planning should follow the patterns proposed for the other four cases.

This discussion of Cases IV and V brings us to the end of our examination of air superiority. We have looked at it from a variety of angles, to grasp as much of it as possible. Each case presents its own special problems, but obvious in every case is the clear mandate to concentrate forces. No simpler—nor more often ignored—principle exists than this one. The commander who concentrates his forces either wins or staves off defeat. The commander who doesn't, loses or wins by accident.

We will see the same principle at play in the next chapters on interdiction and close air support.

Interdiction

Clausewitz said a century and a half ago that combat—battle—was the essence of war. Under some circumstances, it might not be necessary to engage in actual battle. But even in such cases, the threat of war, if not the actuality, determined victory or defeat.[1]

Battle, in its simplest terms, is the clash of armed men on a front.

For armed men to clash, for there to be battle, the men, their weapons, ammunition, food, and information must get to the front. If they are already at the front, then sustaining support, such as reinforcements and materiel, must reach them. The totality of men and equipment moves from its source to the front along lines of communication that can range from primitive trails to complex air routes. In the case of materiel, sources reach all the way back to the raw material from which the materiel was made. In addition to the lines of communication that run from the source to the front, lines of communication also run laterally along the front. Over these lines move troops shifted from one part of the front to another to meet threats or exploit opportunities.

From the earliest recorded times, commanders have sought to place their forces between the enemy and his base. So serious can such an interposition be that during certain periods, notably in the eighteenth century, this act alone, without any battle taking place, often was enough to induce the interdicted side to make peace. Thus, the history of interdiction is as long, and nearly as important, as the history of battle. The advent of the airplane only added a new dimension to this form of warfare.

Many definitions of interdiction exist. Sometimes it is even broken down into subcategories, such as battlefield interdiction. For simplicity, we will consider any operation designed to slow or inhibit the flow of men or materiel from the source to the front, or laterally behind the front, as interdiction. Additionally, as we did in discussing air superiority, we will not make any distinctions between operations directed at the source and those targeted immediately behind the lines.

Thus, an attack on a train carrying iron ore to the smelter is just as much interdiction as destroying a bridge a mile behind the front. Naturally, the time period required for the effect of either to be felt will vary enormously. Even so, both are interdiction and may fit into the commander's theater air campaign.

With the exception of direct attacks on the source of war materiel, the effectiveness of interdiction is tied closely to either the friendly or enemy ground situation. In general, it is most effective when the enemy is under pressure from hostile action or because his own plans demand mobility. To help in visualizing these situations, we will divide ground action into six categories and examine each in detail.

IN RETREAT

The most serious predicament with which a ground force must deal is a retreat under enemy pressure. Under such circumstances, the ground force must slow the enemy pursuit as much as possible to make time for establishing new lines, evacuation, or arrival of reinforcements. Interdiction can buy the needed time. Unfortunately, problems that led to the retreat—especially if it is retreat on a theater scale—probably included loss of air superiority. Under some unusual conditions, however, air superiority may not have been lost and air interdiction may be possible. The American retreat from the Yalu River in Korea in the late fall of 1950 is a good example.

After a very successful offensive, reaching to the banks of the Yalu, MacArthur's 8th Army and Xth Corps encountered a massive Chinese counteroffensive. Badly outnumbered—in some cases by as much as 10-to-1—MacArthur ordered a general retreat. General Lin Piao, the Chinese commander, started his pursuit with the objective of destroying the 8th Army as far north as possible. To accomplish this end, he had to abandon his previous practice of marching only at night and camouflaging his army during the day. When he started moving in the day and driving at night with convoy lights on, to develop sufficient speed to catch the retiring American forces, he exposed himself to American air.

The Americans, who still had air superiority, exploited Lin Piao's exposure to the maximum. American intelligence, using aerial reconnaissance and extensive interviews of captured Chinese, estimated that in December alone air attacks killed and wounded more than 30,000 Chinese soldiers—the equivalent of four to five full divisions. The Chinese were unable to sustain such a high casualty rate and were forced to resume their previous practice of marching at night and hiding during the day.[2] MacArthur's armies escaped virtually intact, with total casualties of less than 13,000 killed and wounded.[3]

Interdiction worked well in this case because the Americans had air superiority and because the Chinese were forced to expose themselves in order to carry out their operations. When we examine the other end of the spectrum, we will see what can happen when a retreating force does not have air superiority and is subject to interdiction.

STATIC DEFENSE AGAINST AN ENEMY OFFENSIVE

The next most serious situation in which ground forces can find themselves is in static defense against an enemy still on the offensive. The early part of the Korean war provides another good example of what air interdiction can accomplish.

By the first part of July 1950, the surprise North Korean attack of the previous month had pushed South Korean forces and American reinforcements to the far south of the Korean peninsula. There, around the port city of Pusan, the allies succeeded in establishing a defensive perimeter—but one they feared might break at any time. Despite the severity of the situation, MacArthur wanted to begin a counteroffensive. He could not do so, however, until significant forces could be brought from the United States. Important to his counteroffensive plans was maintenance of the Pusan perimeter. He decided that it could only be held if his air arm could keep the North Koreans from massing enough men and supplies for a final effort. He opted to use his air to conduct an intensive interdiction campaign. The campaign succeeded and the perimeter held.[4]

This interdiction operation took place under air superiority (although at this point in the war, neither side had large air forces available).

OFFENSIVE OPERATIONS ON BOTH SIDES

The next step on the spectrum from worst to best is the situation in which both sides are roughly equal and attempting offensive operations. Under these circumstances, neither side may have sufficient air superioroity to conduct an effective interdiction campaign. If an interdiction opportunity does present itself, however, it can pay big dividends. The desert battles of 1941 in North Africa provide an interesting example of the case where neither side had air superiority in the immediate battle zone, but where one side, the British, was able to conduct effective interdiction some distance from the actual fighting.

After initial British successes, under General Richard N. O'-Connor, fighting in the desert became inconclusive until Rommel assumed command of Axis forces. By the late fall of 1941, Rommel had driven the British back to the Egyptian border and was poised for a final offensive. While suffering reverses in the ground battle, however, the British had waged an intensive interdiction campaign from Malta against Axis shipping to Libya and Tunisia. The campaign reached its height in the fall of 1941, when British air and naval units succeeded in destroying in September 38.5 percent of all supplies sent to Rommel. In November, the British destroyed 77 percent.[5] Thus, in December,

Rommel was down to 40 tanks, his ammunition stocks were dangerously low, and he was told by the Italians that they had no way to get anything to him for at least another month. He had no option but to retreat from Tobruk and the Egyptian frontier.[6]

After Rommel's reverses in the desert, the German high command belatedly recognized that British interdiction operations from Malta were intolerable; consequently, they mounted a massive air attack on Malta that came close to causing the island's garrison to surrender. British naval and air units no longer were able to operate from the island. The air attacks, which began in December, had an immediate salutary effect; in January 1942, the Germans lost only 20 percent of their shipping.[7] Taking advantage of the neutralization of Malta, they moved sufficient supplies to Rommel to permit him to undertake a major offensive in April.

The interdiction from Malta worked because the British had air superiority over and near it, and because the British in North Africa were maintaining more pressure on Rommel than he could stand. The interdiction effort came to a grinding halt when the Germans seized air superiority over the island. Malta is a classic case illustrating what can happen when things apparently peripheral to the main operation are ignored. Had the Italians taken Malta at the beginning of the war—as they could have— or had the Germans mounted their air attacks on it some months earlier—as they could have—Rommel quite possibly would have prevailed.

OFFENSIVE OPERATIONS AGAINST A STATIC DEFENSE

Let us next examine the condition in which ground forces are intent on launching an offensive against an enemy in static defense. To withstand the attack, the enemy must have sufficient supplies and, except in extraordinary cases, must have the ability to commit reserves and move forces from one point in the line to another. Interdiction can restrict both moves, but experience indicates that its principal benefit is in slowing or stopping the movement of reserves and reinforcements. Two campaigns, the Allied attack on the Gustav Line and the Allied invasion of Normandy, are illustrative.

In the fall of 1943, the Germans in Italy established a fortified defense, the Gustav Line, along the Garigiliano and Rapido rivers on the west and on the Sangro river on the east side of the Italian peninsula. The Allies made attack after attack on the Gustav Line, starting in October 1943. Losses were high and gains were negligible. In an effort to break the German defenses (and minimize casualties), the American air forces began operation "Strangle" in March 1944. Designed to stop the flow of supplies to the Germans, it focused on railroads and roads well north of Rome. In the first week, the Allies cut every railroad in at least two places. Thereafter, they averaged 25 cuts per day. Rail capacity fell from 80,000 tons per day to 4,000, well below what the Germans needed to resist an intensive offense. On 4,000 tons a day, however, the Germans could survive in the absence of Allied ground attack. Thus, they did not withdraw.[8] The next step then was resumption of the ground offensive.

In preparation for the start of the offensive, the US air forces continued operation "Strangle," but shifted the focus to the area immediately behind the German lines to just north of Rome. The Allies launched the ground attack in the middle of May, broke through quickly, and in the 14 days after the initial attack, linked up with the beleaguered beachhead at Anzio. In another 10 days, they took Rome.

In just over three weeks, the Allies, who in the preceding six months had achieved nothing with great losses, now moved 80 miles and forced the Germans into precipitate retreat. Losses were relatively light. Since the ground force ratios did not vary significantly from what they had been during the abortive attacks of the late fall and early winter, air interdiction apparently had done the job—as indeed it had, but not quite in the way that had been anticipated.

Postwar interviews with German commanders, and reviews of German records, indicated that the Germans had sufficient supplies on hand to meet the attack. The northern interdiction had not done too much damage from that standpoint. What made an enormous difference, however, was the German inability to move reserves to the front or to move forces laterally along it. The interdiction campaign had taken such a toll of trucks and trains, and had done so much damage to bridges, railroads, and roads, that the Germans were dependent on foot power and animal transport to move anywhere. Interference before and during the offensive with lateral lines of communication was especially effective.[9]

General Frido von Senger und Etterlin, commander of the XIV *Panzer* Corps during the battle, said that enemy air control created difficulties for the German defenders to move troops laterally as was required. He was only able to move at night. He noted that the commander who could only move during darkness was like a chess player allowed only one move for each three made by his opponent.[10]

The interdiction campaign in Italy succeeded because the Allies had air superiority and were able to keep constant pressure on the lines of communication. It did not, however, force the Germans to retreat. A ground offensive was needed to do that. The offensive succeeded where previously it had failed because the interdiction effort kept the Germans from moving forces needed to plug gaps in the line.

The concept of combining an interdiction campaign with an offensive on the ground is of such importance as to merit another example. The Allied invasion of Normandy was planned in the full knowledge that German forces in northern France would greatly outnumber the invaders. The only way the invasion could succeed was by preventing the movement of reinforcements to the Normandy area. The planners depended on an interdiction campaign to accomplish that end.

The Germans had two basic options for defense against an invasion they knew would come. First, they could put everything they had on the beaches and hope they would be either sufficiently strong everywhere or that they would correctly anticipate the site of the landings. Under this option, the Germans hoped that the Allies would never be able to establish a beachhead. Second, in consonance with German doctrine, they could keep forces in reserve until the main landing was clear. They then could hurl superior forces against the beachhead to destroy it.

Rommel, who had experience trying to move forces when the enemy controlled the air, argued strongly for the first option. Von Rundstedt, who had no significant experience with enemy air, argued just as strongly for the second.

Von Rundstedt won.[11]

The numbers alone would seem to have justified von Rundstedt's position: Under his command, he had a million-and-a-half men organized in 60 divisions.[12] Contrasted to von Rundstedt, the Allies were only able to put ashore a total of 325,000 men in the first week.[13]

The difference was in air power. The Allies had conducted a two-month interdiction campaign before the invasion. On D-day, the Allies flew 14,000 sorties, opposed to 100 the Germans managed to put in the air. (Allied air losses from all causes were 127 aircraft, while German losses were 39.)[14]

The interdiction campaign had two phases. The first phase, begun in the early spring of 1944, was designed to overwhelm the German transport system by destroying railroads, bridges, and rolling stock.[15] The second phase was to prevent the movement of German forces to Normandy after the invasion started. The success of the first phase is evidenced by a report from Colonel Hoffner, officer in charge of railroads in von Rundstedt's area.

In May 1944, he told von Rundstedt that the Germans needed 100 trains a day. The Germans, however, only managed to average 32 trains a day in that same month—down from 60 a day in April. May traffic was only 13 percent of what it had been in January.[16]

The interdiction campaign had crippled the Germans in France before the Allies waded ashore at Normandy and continued to do so after the invasion commenced. Reports of senior German generals attest to the success of the effort: Von Rundstedt said, "The Allied Air Force paralyzed all movement by day, and made it very difficult even at night." Von Kluge, von Rundstedt's successor, communicated, "The enemy's command of the air restricts all movement in terms of both space and time, and renders calculation of time impossible." And Rommel commented, "Our operations in Normandy are tremendously hampered, and in some places even rendered impossible" and "the movement of our troops on the battlefield is almost completely paralysed."[17]

The interdiction campaigns in Normandy and Italy were successful. In the first case, they allowed a landing to succeed that was impossible without them. In the second, they permitted an offensive to succeed when it previously had failed with heavy loss. Two things stand out about these operations: The interdictors had complete air superiority and the defenders were put under enormous pressure by the attackers. Together, the two are powerfully synergistic.

AGAINST A RETREATING ENEMY

Napoleon once observed that no sight is dearer to the soldier than the knapsack of his enemy. Indeed, that sight is even dearer to the modern soldier who has air power to help him in the pursuit. The retreating army is especially vulnerable to interdiction for several reasons.

- First, it has probably lost air superiority.
- Second, by definition, it is in a hurry, and thus less capable of taking elementary precautions against air attack.
- Third, it probably has lost much of whatever ground-based air defense it had.
- Fourth, it may have lost the leadership and discipline that usually provide rational direction.

Taken together, these four factors make a retreating army an ideal target for air action.

The concept of air interdiction against an army in retreat should be clear enough that it does not need many supporting examples. One should be adequate.

After the Allies broke through the Gustav Line in Italy in May 1944, the Germans fell into a rapid retreat. A retreating army is not necessarily a routed army, and the Germans maintained fair order as they pulled back to the north. Despite their relative order, they still lost more than 70,000 men killed and wounded and a great amount of equipment to air action.[18]

Of course, the preceding interdiction campaign, which had made movement to the south difficult, also made retreat to the north difficult. Compare these losses to the losses suffered by the Americans when they retreated from the Yalu River in Korea. The difference comes from the fact that the Germans had lost air superiority, while the Americans had maintained it.

AGAINST SELF-SUFFICIENT FORCES

The last category in the spectrum is the special situation where the enemy force is self sustaining, or nearly so. Guerrillas sometimes are nearly self sustaining in early phases of their war. Even main force units can be almost independent when they are fighting a very low-intensity war, or if they are under no great pressure from their enemy. Obviously, a force that needs little or nothing to exist or fight does not need the kind of supply lines that make air interdiction worthwhile.

Examples are copious and include the Vietnam War before North Vietnamese army forces moved to the south in strength around 1965, the Mau Mau uprising in Kenya, and partisan operations in the Balkans in World War II. The mere fact that a thing called air interdiction exists does not mean that it is appropriate for all conflicts. In the examples cited, it wasn't, and any attempt to use it probably represented effort that should have been devoted to something more productive.

We have seen what interdiction can do and where it is most effective—when the pressure is on the enemy and he needs to move major forces and equipment quickly, such as during a retreat, a pursuit, or a defense against a determined offense.

Now, we must look at where to interdict. A simple three-level categorization based on relative distance from the front gives us a framework for analysis and planning. Interdiction can be close, intermediate, or distant.

We may define *distant interdiction* as an attack against the source of men and materiel, or, in the case of a warring party that has no industry, the ports or airfields where materiel provided from outside enters the country.

Intermediate interdiction occurs somewhere between the source and the front.

Close interdiction is interdiction in that area along the front where lateral movement takes place.

It is possible to concentrate on one, or on all together if the circumstances are proper and enough air resources are available for proper concentration on each area. Again, examples from past conflicts will give the reader an idea of the problems and opportunities associated with each.

Distant interdiction has the capability of producing the most decisive outcomes affecting the whole theater—or even theaters—but it also has attached to it the greatest time lags between attack and discernible result at the front. For instance, if every oil refinery in the world blew up tomorrow, oil-based industry and transportation wouldn't be forced to shut down the following day. In some cases, they could continue to operate for weeks or even months. Eventually, though, they would stop if the refineries were not rebuilt. The lesson for the air and theater commander is that a delay always exists between cause and effect. If the commander is sure that the war will be decided before an effect can be felt from a given action, then it is pointless to waste resources on carrying it out. He needs to be very careful in this assessment, however, for wars are inevitably much longer or much shorter than anyone expects.

The classic examples of *distant interdiction* are the operations conducted against Japan and Germany in World War II. Since these operations already have been discussed in some detail, it should suffice to recall that four months of bombing attacks against German petroleum facilities reduced aviation gas production by 98 percent. By the end of the war, German tanks, if they were lucky, got only enough fuel for a single attack.[19]

A modern society, let alone a modern army and air force, cannot operate without fuel. After a success of the magnitude discussed above, one need only wait (while maintaining a degree of pressure) for the enemy to callpase. Barring utter fanaticism, which was unexpectedly stronger in Germany than in Japan, the enemy will quit.

Intermediate interdiction also has a time lag associated with it, but one that probably will be less than that for distant interdiction. It has been most suitable in preparing for future operations. For example, the attacks on rail lines in Italy and France did not pay off until the offensive began against the Gustav Line and until the invasion fleet landed at Normandy. Then the payoff was great.

A similar situation existed in the Southwest Pacific, when the American air force caught a huge Japanese convoy moving troops and supplies from Rabaul to the Huon Peninsula. The convoy was destroyed and a whole division lost.[20] Its loss did not have an immediate effect on the battle that was taking place a couple of hundred miles to the east. But its effects were felt months later, when MacArthur moved against the Huon bases that the convoy had intended to build up. MacArthur also conducted a continuous campaign against barge traffic that tried to run from Hollandia and Wewak to intermediate locations the Japanese thought would be strong points that would have to be reduced before the Americans advanced. Although MacArthur bypassed most of these bases, the long and effective interdiction campaign had succeeded in isolating them from resupply of food, spare parts, and medicine.[21]

Thus, the bases he did attack were considerably weakened and were not able to put up the resistance they otherwise might have.

The last of the categories of interdiction—*close interdiction*—seems most useful when a battle is in progress. We have seen how it played such an important role just prior to and during the May 1944 attacks on the Gustav Line, and even more so at Normandy. Another, more recent example occurred during the 1973 Arab-Israeli war.

On Sunday, 7 October 1973, the Syrians committed their armor reserves on the Golan Heights. Three hundred tanks, commanded by the Syrian President's brother, drove to within five miles of the Benot Yacov bridge. Nothing stood between them and the opportunity to debouch onto the plains below Go-

U.S Air Force Photo

A KC-135R tanker (bottom) gets ready to refuel an F-16 *Fighting Falcon,* a highly maneuverable fighter capable of speeds in excess of Mach 2 to support the air campaign. A multi-barrel 20mm cannon is mounted in the fuselage, along with weapons or tanks on four underwing pylons and air-to-air missiles on the wingtips.

lan except a handful of reservists just arriving at the front. But, just as a serious setback to the Israelis seemed imminent, the Syrian advance "ran out of steam." As it turned out, the Syrians had run out of gas and ammunition because "the Israeli Air Force had destroyed it." The previous night, the Israelis had conducted interdiction operations just behind the front against the convoys of Syrian trucks carrying ammunition and fuel. This operation, conducted in lieu of close air support despite the desperate ground situation, had a major impact on the battle.[22]

Interdiction is a powerful tool in the hands of the joint and air commander, a tool he can use either as part of a potentially war-winning campaign—distant interdiction against the source—or as part of the ground campaign. It will be effective to some degree in almost all situations. But when it is used in support of ground operations, it is most effective when the enemy is under, or is about to be put under, severe pressure. However, some potential problems must be considered.

All of the successful interdiction campaigns we have discussed have been sustained, concentrated efforts. (The Israeli interdiction of Syrian supplies on the Golan was short, but compared to the total length of the war, it probably wasn't much different from the operations in Italy prior to the attack on the Gustav Line.) It is futile to believe that one or two missions by a handful of planes are going to accomplish anything lasting. Like everything else we have discussed, mass and concentration are essential.

Interdiction operations are going to lead to loss of aircraft and flyers; thus, it is necessary to ensure that something useful is gained for the loss. One modern aircraft and a highly trained pilot probably are too high a price to pay for one old truck, loaded with rice, driven by a private.

Interdiction operations should not be done at the expense of something more important. That something more important almost certainly will be air superiority. A ground commander will demand interdiction in many instances before air superiority has been won. Interdiction missions, except under unusual circumstances, when the benefit clearly outweighs the risk, should not be attempted in the absence of air superiority. A commander does so at his peril, for he is likely to jeopardize his chances of ever winning it. We have seen very clearly what fighting without air superiority is like.

A possible compromise, however, may be possible between the demand for interdiction in support of ground operations and the need to achieve air superiority as quickly as possible. Simply, some targets may support both. A prime example is fuel.

The same fuel probably will not be used for an enemy fighter and an enemy tank, but it almost certainly will be transported and managed by the same net. Attacks on petroleum consequently will serve both ends. Attacks on the transportation net may do the same. Additionally, the enemy must respond with fighters to strikes against his petroleum or transport net. When they do respond, fighters escorting interdiction aircraft can attack and destroy enemy aircraft in the air—keeping in mind that the most expensive place to take out enemy air is in the air.

Nevertheless, if the previously discussed precepts of mass and numerical superiority are observed, these operations can be doubly profitable.

With this observation, we conclude our discussion of interdiction. The next chapter will deal with the use of aircraft to strike enemy troops directly, at the front, in coordination with friendly ground forces.

7

Close Air Support

Close air support has been around since the early days of the aircraft in World War I. Although it has gone under other names, such as army cooperation, close cooperation, and ground support, every major country with an air force has tried it in some form. The Russians in World War II, for example, did little else and treated their air force as mobile artillery.[1]

At the other extreme, the Israelis, according to a recent commander of the Israeli Defense Forces, have shied away from close air support as a mission. The Israelis believe that teaching pilots to identify friends is too difficult; in any event, the mission is to identify and strike enemies.[2]

The Germans in World War II, at least after the Battle of Britain, tended to be more like the Russians, while the Americans tended more to the Israeli position.

Close air support can look like interdiction, and vice versa. To help reduce the confusion, finding common areas of agreement and disagreement is useful. An air attack on enemy forces crossing the wire 50 yards from friendly troops, and controlled at least indirectly by the concerned ground commander, certainly is close air support. Just about everyone will agree that air attack on enemy troops within rifle range of friendly forces also is considered close air support. Similarly, just about everyone would agree that air attack on a tank factory is not. Clearly, substantial room is left between these two extremes.

Procedure might provide a basis for identification. One could say that aerial attack on anything within range of artillery would be close air support, because attacks on targets in that zone would require coordination with the ground commander. This answer is not completely satisfactory, however, because targets within artillery range—say 20 miles—possibly may have nothing to do with the current ground situation. Targets in this category might include airfields on which air combat fighters are located, radars used as part of the enemy early-warning system, or even enemy troops that happened to be moving laterally across that sector of the front. Procedure not only does not solve the problem it expands the area of confusion. Another approach is needed.

Let us define close air support as any air operation that theoretically could and would be done by ground forces on their own, if sufficient troops or artillery were available. Under this definition, air strikes on troops crossing the wire certainly would fit the category. Aerial bombardment of the enemy line, preparatory to an offensive, also would fit, because artillery could do that job. Using air to hold a flank fits under the rubric of close air support, because an extra division or corps could be assigned flank-holding duties. Aerial attacks on enemy troops moving laterally across the front does not fit, however, because ground forces have no realistic way to deal with that kind of action (other than perhaps harassing fire of some sort). If an air action does not fall within this definition, for our purposes, it will be either interdiction or air superiority.

This definition may or may not agree with the definition currently in use in any particular army or air force. It is not important that it does. What is important is that air—and ground—commanders go through a mental exercise to differentiate between close air support and all other air operations. It is of more than theoretical importance that they do so.

The ground commander must play a key role in determining where close air support will be employed. Naturally, ground commanders tend to concentrate on their immediate job, which is to advance on the ground or to prevent the enemy from doing so. If the definition of close air support becomes too broad, then these commanders in effect exercise control over great parts of the air forces. In fact, this very thing happened to the *Luftwaffe* on the Russian front, where the army monopolized air assets. Only in 1944, when it was too late, did the army recognize that interdiction would have been far more productive than close air support.[3]

It could be argued that the preceding discussion implies parochialism. It does, but not in a pejorative sense. Ground commanders and staff, at least at army level and below, tend to be vitally concerned with their immediate front—as they should be. Similarly, political leaders at home can look at the movement of lines on a map and judge progress or lack thereof at the front. The measurable, obvious indications grip everyone, and the pressure to move lines on the map becomes inexorable.

Conversely, it is quite difficult during the course of a way to display the effect of interdiction operations that perhaps prevented enemy divisions from joining the battle. Almost until the end, it is hard to display and comprehend the impact of strikes against the enemy's petroleum industry or transport system.

The air force officer is trained to look at a different front. He thinks in terms of air superiority, soft spots in the petroleum system, transport net vulnerabilities, and opportunities to interdict divisions weeks away from commitment. Because of the mobility of his medium, he tends to look at greater expanses of space and time. In the process, however, he may relegate the movement of lines to a subsidiary part of the big picture. He may lose touch with the tangible flesh and blood of the front as he surveys his own domain, where casualties may be as high as on the ground, but where the bodies are rarely seen—and where no lines are drawn on a map.

Powerful forces are pulling the ground commander one way and the air commander another. If a rational air campaign is to be carried out—whether for air superiority or interdiction, or both—an air force must have freedom to do it. The air campaign, under some circumstances, may be far more important that the ground campaign. That never will come to be, however, if the definition of close air support, or the importance attached to it, becomes so inclusive that the ground commander exercises effective control over large parts of the air force. As a minimum, the theater commander should decide which campaigns are to be emphasized. To make these decisions, he needs candid advice from ground and air component commanders. They in turn must each have a thorough, unambiguous doctrinal understanding of how their respective Services can contribute to winning the war.

Having disposed of the thorny doctrinal problem of defining close air support, we now can look at how it can or should be used.

First, by his very nature, the soldier on the ground will find close air support useful in almost every conceivable situation, from pursuit to retreat. If it were available, the man on the ground would like to see air precede his every move. No air force has yet been large enough, even when totally subordinated to the army, to provide that level of service. The one possible exception may be the anomalous Vietnam War, where the American air force took part in most ground engagements.

Given that close air support is desired by everyone, but cannot be provided to all, how can this limited resource best be used? The answer is inherent in the definition we proposed and in the nature of the aircraft. We suggested that close air support was a substitute for something that could be done with more divisions or more artillery if they were available—and if they could get to the battle in time. Left hanging is the question of when the extra division or artillery should be employed. The answer lies in the concept of the operational reserve.

We will discuss the concept of reserves in detail in the next chapter; for now, suffice it to say that the operational-level reserve is normally committed to exploit a great opportunity—either positive or negative. That is, commitment of the reserve is appropriate if doing so will allow a commander to make or extend a breakthrough (a positive opportunity) or will allow him to preempt or to stop an enemy breakthrough (a negative opportunity).

If we think of close air support in terms analogous to the operational ground reserve, we tend to put proper value on a scarce and valuable commodity. We put it in terms both the airman and the soldier can understand. We also make it easier to comprehend that close air support, like the operational reserve, is something to be used quickly and decisively. It is a shock weapon that is most effective when concentrated in space and time.

The speed and mobility of the aircraft facilitate concentration and employment. The two characteristics also mean that aircraft participating in other missions can, under the right circumstances, be redirected in a matter of hours to close air support if the opportunity for appropriate employment presents itself. Thus, the options for employment are more numerous than for a ground reserve that may take days to build and commit.

On the other hand, a single airplane normally cannot stay on station very long, and there are rarely enough of them to maintain an around-the-clock coverage.

We now have two ideas for where to use close air support: where an operational-level commander would want to employ his own operational reserve and where bursts of power—as opposed to the long-term power of ground forces—are indicated. Commanders have historically used their operational ground reserve to break through the enemy lines, to prevent an enemy breakthrough, or to cover a flank. Close air support has accomplished all these missions. Let us look at some examples.

In the 1940 offensive in France, one of the first problems confronting the Germans was how to cross the Meuse River with three divisions opposed by three French divisions dug in on the opposite bank. An attack by *Stuka* dive bombers offered the key. But the question then arose as to whether one massive attack, as was consistent with *Luftwaffe* doctrine, would do, or a continuous attack would be carried out, as requested by the ground commander, General Guderian. Guderian explained that he needed to keep the enemy down while he made his initial crossings. A single attack would not accomplish that end. The air force then agreed to provide him with a stream of *Stukas*. The air attack took place, three divisions crossed the river to overwhelm three defending divisions, and a breakthrough was underway.[4]

One year later, the Germans again used air to spearhead a breakthrough—this time on the Eastern Front. On 23 August

1941, the *Luftwaffe's* VIII Corps (its dedicated close support unit) flew 1,600 sorties to open a way for a 60-kilometer advance by Wietersheim's *Panzer* Corps. During this massive attack, the *Luftwaffe* lost only three aircraft, while destroying more than 90 Russian machines.[5]

The Americans also used air extensively for breakthrough operations. Normandy was the greatest such effort ever mounted. We already have discussed the role of air in interdicting movements of German forces to the beachhead area. But on the day itself many sorties were flown directly against German defensive positions. One particular mission is worth detailing, because it illustrates the use of air to do something more than make brute force holes in a line. After learning from "Ultra" intercepts the location of the *Panzer* Group West headquarters, Allied aircraft struck the headquarters in force. They destroyed this key command structure, which had the very critical job of coordinating armor movements, and killed a number of staff officers, including the chief of staff.[6]

In the Pacific theater, massive air attacks preceded amphibious landings in virtually every case. The Strategic Bombing Survey concluded that "sustained air preparation for landing operations against well defended positions materially reduced the casualty rate."[7] Many factors affected the outcome of island invasions. But, remarkably, Japanese casualties tended to be 10 times higher than those of the attackers—an outcome due in part to American air superiority and concomitant air attack.

Almost a quarter century later, in the 1967 Six Day War, the Israelis used their air to blast through fortified positions on the Syrian controlled ridge running north from the Sea of Galilee. Israeli armor poured through the gap to begin a pursuit that would carry it to within 25 miles of Damascus.[8]

In addition to opening the way for breakthrough operations, air also has provided the ground commander with protection for his flanks. The first significant use of air for this purpose took place in France in 1940, when the *Luftwaffe* was charged with covering the flanks of the army's deep armor penetrations.[9] It performed a similar role by turning back a massive Russian attack on the right flank of the 4th *Panzer* Army, which was moving north for the offensive against the Kursk salient[10] in the summer of 1943.[11]

A year later, in France, the tables were turned when General Patton gave the XIX Tactical Air Command (TAC) the job of protecting his exposed flank along the Loire River as he raced to the east. So successful was the operation that the commander of German forces south of the Loire requested that the XIX TAC commander be present when he surrendered his command and 20,000 German troops.[12]

On the other side of the world, MacArthur used Kenney's air force to guard his flanks. In September 1944, MacArthur needed another airfield for his move against Leyte. He chose the lightly occupied island of Morotai, instead of the heavily defended main island of Halmahera, less than 60 miles to the south. When MacArthur moved on to Leyte, he expected the air force to keep the 30,000 Japanese troops on Halmahera away from Morotai.[13]

On an as yet unequalled scale, MacArthur made the air force responsible for protecting the flank of one of history's longest operational penetrations: MacArthur had left strong Japanese forces intact on New Guinea, on Halmahera, in the Netherlands East Indies, and on Mindanao.[14]

Our next illustration comes from China. There, the Japanese armies had made a series of attempts to take the key cities of Kumming and Chunking. They all failed. After the war, the Japanese ground commanders reported that at least 75 percent of the resistance they encountered had come from air attacks mounted by 14th Air Force. The 14th, consuming supplies that would have supplied about a division of ground troops, kept a 500,000-man army from reaching its objectives.[15] This particular example could be interpreted as more interdiction than close air support, but it fits our definition of something that could be done by the operational reserve if it were available to do it.

At the end of 1942, General Paulus's 6th Army found itself in dire straits after the Russians had encircled it at Stalingrad. Some evidence exists that the Russians had waited for weather bad enough to restrict *Luftwaffe* activity before launching the first part of their offensive against Paulus.[16] By doing so, they achieved a degree of defensive air superiority. When the bad weather broke, however, the small *Luftwaffe* force (less than two dozen fighters and bombers), flying from fields within the encirclement, managed on several occasions to turn back substantial armored thrusts.[17] It was, of course, unable to hold the Russians off indefinitely.

Six months after Stalingrad, the *Luftwaffe* played a key role in the action around Kursk and Orel. The Germans were trying to cut off and encircle Russian forces in the Kursk salient, while the Russians were trying to do the same thing to the Germans in the Orel salient just to the north of Kursk. For lack of reserves, the Germans had to abandon the Kursk offensive to defend the Orel area. Key to the defense was the *Luftwaffe*. Every aircraft available on the eastern front was concentrated to meet the big Russian attack, which came on 19 July 1943. General Model, the German commander at Orel, gave "full credit" to the *Luftwaffe* for stopping the offensive. He said that the German success was "from air alone."[18]

Our final example of close air support in the defense comes from Vietnam. American commanders deliberately enticed North Vietnamese attacks on strongpoints that could be supported by close air operations. *Khe Sanh* provides a dramatic example. More than two divisions of North Vietnamese, consisting of 15,000 to 20,000 men, besieged an emplacement manned by 6,000 marines. The North Vietnamese, with incredible tenacity and bravery, made attack after attack on the Americans. Despite their numerical superiority, however, they were unable to prevail against the 350 fighter and 60 bomber sorties that flew against them every day for two months. They finally were forced to lift the siege in March 1968 and fall back with terrible casualties.[19]

American losses on *Khe Sanh* were comparatively light. More than a decade before *Khe Sanh*, the North Vietnamese had won an impressive victory over the French in almost identical circumstances at *Dien Bien Phu*. The difference between the two battles was simple: The United States had massive air power; the French did not.

USAF Photographic Collection, National Air & Space Museum, Smithsonian Institution

One of many B-26 Martin *Marauders* of the 9th Air Force over the coast of France during the early morning gives cover to landing craft on the sandy beaches below during the D-Day invasion of France early in June 1944.

As we have seen, close air support can do a lot for the ground commander. It is not, however, without its problems. One very important deficiency is the inability of close air support to operate when the weather is bad. The commander who counts on close air may be badly shocked if it is not available. Conversely, the commander who is trying to operate without significant air support may be able to execute a movement in bad weather that would be impossible in good weather, when enemy air could strike him repeatedly. Remember the Russian exploitation of bad weather for the Stalingrad counteroffensive. Perhaps the Russian commanders had gotten the idea after seeing the decrease in German speed of movement when bad weather in the fall of 1941 forced the *Luftwaffe* to reduce its daily sortie rate from 1,000 in September to 269 on 9 October.[20]

In retrospect, it is difficult to conceive of Germany's 1944 counteroffensive in the *Ardennes* getting started at all had not bad weather made it almost impossible for Allied air to operate against it. At some point in the future, conducting close air support in bad weather may be possible. Until that day, however, air and ground commanders must be aware that weather can have a significant effect on their plans.

Close air support may not always be available, even in the presence of a large air force. First, the need for it may be higher on some other part of the front. Second, and crucial for the ground commander to understand, other missions, such as air superiority or interdiction, may have higher theater priority. As discussed in chapter 3, air superiority is a theater necessity. Until it is won, any effort not contributing to it is diversionary and should only be undertaken in emergency situations, when the risks and rewards have been carefully considered. As an example, Kenney refused air support to the Australians attacking Salamaua; he needed to concentrate his forces against Rabaul to win the air superiority that was a sine qua non for the overall campaign.[21]

The air component commander and the theater commander must consider the cost of providing close air support. It almost always will cost something, even if that something is only lost opportunity. We saw at the beginning of this chapter how the German army realized late in the war that it had misused the air force by committing it so heavily to close support. The Americans and British won the war, but they also debated the efficacy of close air support.

The most notable debate was over bomber support for the Normandy invasion. The air force had argued strenuously that the big bombers should continue their efforts against the German homeland. The Combined Chiefs overruled the air arguments and ordered maximum bomber operations in France. Normandy was a big success and wouldn't have been without heavy air support. Nevertheless, the cost for diverting the bombers was a three-month delay in attacks on the German petroleum industry. Would the war had ended sooner if the Ger-

U.S. Air Force Photo

An A-10 *Thunderbolt II* banks to the right. The aircraft can carry four Maverick Scene Magnification missiles and an electronic countermeasures pod. Known as the "Tank Buster," the A-10 is a close-support attack aircraft developed as a special ground-attack aircraft.

mans had run out of fuel for their tanks and aircraft three months before they did?

Obviously, this question can't be answered, nor is it necessary to answer it. The point is that the Allies paid a definite price for diverting air away from Germany. Was the price right? It may well have been, but the price was there just as it will be there in future conflicts. Commanders must decide what they want to pay.

Like interdiction—and the operational reserve—close support seems to work best when the ground situation is dynamic. Close support has the capability to make holes that can be exploited offensively, and it has the capability to do serious damage to enemy offensives. It does not seem to work as well under relatively static circumstances.

This chapter has examined what close air support can do for the ground commander. It also has attempted to show that close air support is not without problems. A price is attached to it. Put in proper perspective, and used appropriately, it can make a big difference in the ground situation. It can provide the ground commander the wherewithal to do things he couldn't do without far more troops and artillery—and very mobile ones at that.

To this point in the book, we have covered the three traditional combat missions for air—air superiority, interdiction, and close air support.

Before attempting to put it all together for campaign planning, we need to consider one other topic that is generally not considered in connection with air operations. In the next chapter, we will look at *reserves* to see how they might affect the air war.

Reserves

This chapter is about operational reserves—a concept that has received little attention in discussions of air operations. The fact that it receives little attention indicates one of two things: Either the conventional wisdom is correct, in assuming that the subject of "reserves" is not applicable to air war, or that it has been ignored mistakenly and is in fact quite important.

Our hypothesis is the latter.

On 16 May 1940, Winston Churchill made a desperate trip to Paris, where he asked the French high command overseeing the hasty retreat of its forces in front of the German offensive through the Ardennes, "*Ou est la masse de manoeuvre?*" ("Where are the reserves?") The answer was, "*Il n'y a aucune!*" ("There are none!")[1]

Just months after the disaster in France, Churchill was in the command post of Number 11 Group watching the progress of the Royal Air Force (RAF) defense against the greatest German raid to that date on London. He had just seen Air Vice Marshal Park commit his last six squadrons to the battle and simultaneously call on his neighboring commander, Air Vice Marshal Leigh-Mallory for all his forces. Churchill asked Park what else he had in reserve. Park answered, "Nothing." All his forces had been committed.[2]

RESERVES MAY HELP BETTER THE ODDS

The theory of reserves is not easy to grasp, especially on an emotional level. We are inclined to feel that a unit not committed to the battle somehow is not pulling its weight. We tend to think of Henry V's injunction of, "Once more unto the breach, dear friends, or close up the wall with our English dead." We think in terms of gathering our strength and charging the objective with everything we have. We accept, perhaps on a visceral level, the theory of concentration and mass, and interpret this theory to mean all of our resources.

We calculate ratios and, never quite comfortable with our superiority (or lack thereof), want to make the odds better. We add more forces in the belief that we are increasing our chance of success or decreasing our chance of failure. In a certain sense, none of these thoughts is entirely wrong; in fact, in a perfectly predictable world, each might be entirely right. But if the world were perfectly predictable, war would never happen. The antagonists, knowing the outcome by virtue of mathematical analysis, would sign the armistice terms before the first bullet flew. War is, of course, an intensely human activity. It defies prediction. That reason is a key to why reserves came to be so important for land warfare.

Clausewitz spoke of the fog of war, the friction of war, and the uncertainty of war. Nothing can eliminate these hindrances to perfect action, but reserves can ameliorate their negative consequences in at least two major ways. First, they provide a commander the wherewithal to exploit an error or failing by the enemy. He can pour into the battle masses of fresh troops who have the potential to break remaining enemy resistance and force a retreat or rout. On the other hand, reserves can be thrown against an enemy attempt to exploit a commander's own error. The arrival of strong, fresh forces may break the enemy attack and restore the line—whether actual on the ground or conceptual in the air.

The point could be made that a sufficiently smart commander ought to be able to anticipate either of these situations, and put requisite forces into action at the start of the battle. A certain theoretical validity can be given to this argument—although the uncertainties of war make the move from theory to practice virtually impossible. What the argument ignores is that the introduction of reserves in effect creates an entirely new battle, or at least a distinctive new phase.

Up to the time when reserves enter the fray, the opponent is dealing with relatively known quantities doing relatively obvious things. Reserves immediately increase the uncertainty factor by dramatically changing the quantitative and manuever equations. In other words, the impact of the sudden appearance of a new division on the battlefield is entirely different from the impact that same division would have made had it been on the line from the onset of action.

Fog and friction are hindrances to one's own action, but to the extent that they can be inflicted on the enemy commander, they become allies. Reserves can help in this process because their mere existence means the enemy commander must take their possible employment into account. Since he doesn't know where they may be committed—or whether a given commitment is a feint or real—he must spend time thinking about them. And he probably will have to deploy his own forces somewhat differently than he would if no reserves faced him.

Subsumed under the broad concept of reserves are some rather elusive principles on how they should be used. Elusive is the only applicable word, for they amount to such useful exhortations as "do the right thing!" A noble expression to be sure, but one difficult to enact. One of the more popular, and actually more valuable, principles applied to reserves is the injunction against committing them piecemeal. This principle has been taught in military schools for centuries—and it has been ignored time after time with normally disastrous results.

Explaining why reserves should not be committed piecemeal is not too difficult: Reserves seem to be most valuable when their appearance shocks enemy troops and commanders. Actually, the mental shock to the enemy may be more important than the physical effect of the reserves. A physical effect, though, can be explained by the principles of momentum. Momentum is a product of mass and velocity. Assuming a constant velocity, the momentum, the force with which something strikes, will be in direct proportion to its mass.

Combining the moral and the physical, we can see why committing reserves piecemeal makes little sense. Piecemeal commitments, because they are incremental, lose much of their ability to induce confusion and fear in the mind of the enemy. Adapting to a gradual change in a situation is far easier than to have to adapt to a sudden and massive change. Second, its potential momentum, and thus its physical shock power, is reduced proportionally to the division of the reserve into small parts.

The next two principles for the use of reserves are simple to state: Don't commit too soon, and don't commit too late.

Clearly, determining what is too early and what is too late must be a highly subjective process. It may even be a work of sheer artistry or genius. The idea is that any battle—or even an entire war—has certain points where the situation is precarious enough, or close enough to equilibrium, that the application of a new force will have an effect all out of proportion to its comparative size. Judging that point or moment is not easy. Confederate General Robert E. Lee thought it had arrived when he sent General George E. Pickett across a mile of open terrain on the third day of the Battle of Gettysburg. Lee guessed wrong—and lost a battle and doomed his cause.

Conversely, Lin Piao made a perfect decision on committing the Chinese—in the sense that China constituted a strategic reserve for the North Koreans—against MacArthur. A few weeks earlier would have required extended lines of communication against a still concentrated American force, while weeks, or perhaps even days, later would have found MacArthur dug in and massed on the south banks of the Yalu.

Timing is everything.

The examples cited and the analogies drawn have so far been from land war. Are the principles applicable to air war? In very broad, theoretical terms one would conclude a priori that they must be. Certainly, the minds of enemy airmen and their commanders are subject to shock, just as are the minds of ground soldiers. Likewise, the physical principle of momentum applies as much in the air as it does on the land. The question is, however, whether the theory translates to practice.

SORTIE NOT FLOWN IS NOT A SORTIE LOST

Flyers have a saying that notes graphically how useless is runway behind them, once their aircraft has touched down for landing. A similar view suggests that a sortie not flown is a sortie forever lost. A general feeling exists that aircraft are to be flown as frequently as maintenance requirements allow, and that a target of some sort will be there for each sortie. These ideas produce a general belief that the concept of reserves does not apply to air operations. In fact, few historical instances exist where air was consciously kept in reserve.

Two exceptions, however, are fascinating. One actually took place during the Battle of Britain, and the other was planned, but never occurred, in the air battle over Germany. Let us review both.

The Battle of Britain began on 8 August 1940, when the Germans opened a campaign in earnest to seize air superiority over Britain to prepare the way for an invasion.[3] The fighter forces opposing each other were roughly equal, but for the British fighters, the German bombers were the target. The Germans had about as many bombers as the British had fighters so, in total, the *Luftwaffe* outnumbered the Royal Air Force (RAF) 2 to 1.[4] (This ratio ignores the British bomber command which, although two-thirds the size of its *Luftwaffe* counterpart, did not make any serious attacks on German fighter bases and thus played no direct role in the Battle of Britain.)

Despite the numerical superiority of the Germans, the British commander, Air Marshal Dowding, kept about a third of his fighter forces away from the battle zone, where they were not subject to attack. Neither could they participate in the war. The British maintained a reserve even during what Churchill called Britain's darkest hour. On top of this, Dowding's two subordinate commanders, Park and Leigh-Mallory, maintained their own reserves.[5] These reserves, Dowding's operational reserve and the sector commanders' tactical reserve, played a key, if not decisive, part in winning the battle for Britain and averting an invasion.

Until the first week in September, the Germans conducted an amateurish attack on Britain that was characterized by shifting objectives from fighter bases to production to shipping. They had failed to follow up on the first day's strikes against British radar. They tried escorted and unescorted bombing missions. When their fighters did accompany the bombers, they used close escort tactics. They frequently failed to concentrate or to coordinate raids on the same day. Despite all these errors, they were making progress. The RAF was beginning to suffer badly.[6]

Then, at the end of the first week in September, Hermann Goering, Nazi Germany's air chief, made a momentous decision, which Hitler fully endorsed. The British, some days earlier, had carried out a militarily insignificant raid on Berlin. An infuriated Hitler agreed to shift the focus of his attack on Britain from military targets to London. Many in the *Luftwaffe* were pleased with the decision that they had been urging for the past two weeks. They thought that it would force the RAF into a decisive air battle.[7]

They neglected, however, the other side of the coin: By concentrating the *Luftwaffe* on London, pressure on fighter fields would subside and, more importantly, the British themselves could concentrate against the *Luftwaffe* more effectively.

The Germans finally decided that the time was right for the final blow. Every aircraft that could be made serviceable was to participate in a coup de grace planned for 15 September. The British, in part through *Ultra* intercepts, knew that the Germans planned their big thrust for the 15th. Consequently, Dowding used his operational reserve to bring every fighter unit in Park's and Leigh-Mallory's sectors up to strength. He also put fewer than the usual number of fighters in the air on 14 September, leading the Germans to think they were winning, and also giving Fighter Command a chance to prepare for the next day's action.[8]

On 15 September, the British met the Germans in mass. Park committed his tactical reserve of six squadrons and asked that Leigh-Mallory bare his sector in order to send all of his forces south to hit the Germans. Leigh-Mallory not only did so, but some of his forces attacked in full wing formation.

As a result of the British use of reserves (and mass), the Germans suffered such heavy losses that they concluded the RAF was so far from being beaten that the *Luftwaffe* did not have the time or resources to do the job. Within days, they shifted to useless night bombings of London.[9] Commitment of air reserves had won the Battle of Britain, had precluded a German amphibious invasion, and in many ways had turned the tide of the war.

Three years after the Battle of Britain, the American daylight bombing campaign was beginning to cause the Germans problems. As previously noted, the German commander of fighter forces, Adolf Galland, had decided that having a 3- or 4-to-1 advantage over the attacking bombers was necessary, to be able to inflict military significant losses on them. (This decision was made before American bombers had fighter escorts for missions over Germany!)

Because of requirements on other fronts, *Luftwaffe* fighter strength in Germany was too low to achieve ratios of fighters-to-bombers of much more than 1-to-1, and that only occasionally. Galland on three occasions proposed to Goering and Hitler that new production and newly trained pilots be withheld from the defense of Germany until sufficient strength was on hand to meet the Americans with at least 3-to-1 superiority. His first plea, in the spring of 1943, included a suggestion that fighters be withdrawn from other fronts; his plea was rejected on the basis that it might lead to lost ground in Russia or in North Africa.[10]

Galland made his next request the following spring. He proposed pulling all fighters out of France to concentrate them against the American bombers. Goering refused again, this time because of the expected invasion.[11]

During the summer of 1944, American bomber attacks became intolerable. Obviously, to men like Galland and Albert Speer, organizer of war production and forced labor for Nazi Germany, the war would be lost if something were not done about the bombers. Therefore, Galland again proposed building a huge reserve for commitment on a single day. He intended to assemble and train this force to put 2,000 fighters in the air on the first day in late fall that the weather was suitable. His objective was to destroy 400 to 500 bombers, with a probable loss of about the same number of German fighters. He believed that one or two days of such losses would make the Americans stop the bomber campaign for an extended period. During that time, Galland hoped his country would have sufficient respite to bring the new jet fighter into service, to restore aviation fuel production, and to train enough new fighter pilots that the Americans would face overwhelming odds when they resumed the campaign.[12]

Would Galland's plan have worked? There is a high probability that it would have.

In the fall of 1943, the American bomber offensive was called off after the 12 to 16 percent losses suffered in October. A loss of 400 to 500 bombers would have been equivalent to a 25 to 50 percent loss rate. Such losses would have seriously damaged virtually every participating unit. The shock itself, after a long period of declining losses, would have been dramatic to flyer and commander alike. The proposed ratios of 4-to-1 would almost certainly have been adequate to accomplish the job. Why didn't it work? Very simply, Hitler disbanded the reserve and committed it in support of the 1944 *Ardennes* counteroffensive.[13]

There, operating in a strange environment in bad weather, flying missions for which it had not trained, and committed against dispersed targets of only tactical importance, the reserve disintegrated. And so was wasted Germany's last hope to avoid destruction.

These two examples from World War II would seem to suggest that air reserves can be of extraordinary importance. Also, they show that the theory of reserves can be applicable to air operations. Do they mean that there must always be an operational or strategic reserve? The answer to that is not clear. The US Air Force has never kept a reserve (although in a sense, its production capacity gave it a strategic reserve in World War II). One could say that its unbroken string of victories since 1943 suggests that it did not need a reserve. Of course, in every conflict since then it has been on the offensive (even if highly circumscribed in Korea and Vietnam), with overwhelming numerical superiority. The two air forces in our examples were on the defensive and were numerically inferior. Reverting to our theoretical discussion, we recall that reserves seem to be most useful when the situation is unstable and susceptible to being unbalanced by the addition of a new force.

These observations would lead us to suspect that air reserves then are most needed when the enemy is equal or somewhat stronger than oneself.

In addition to the theory and examples cited above, some work done with computer war games suggests the utility of air reserves. A brief explanation of one game will help. The simulation started with an enemy numerically superior on the ground and in the air making attacks across a broad front, but more strongly on some areas of the front than on others. The first step of the game was to see how close air support sorties should be apportioned. It turned out, as an exponent of concentration might have expected, that putting the majority of close air support against a particular enemy thrust was far more beneficial than spreading it out over the entire line.

The next step was to examine the possibility that close air support might be more effective on day two or three—or even later—than it was on day one. Before detailing that trial, it is necessary to discuss the initial assumptions on how close air support would be used.

The standard assumption is that as many close air support sorties should be flown as possible on day one. Thereafter, a decreasing number will be flown because of attrition. On a graph, this system would show up as a descending line, starting at a high number on day one and falling to a relatively low number at so many days into the future. In real life, the line might not be straight; rather, it might be a descending sine wave reflecting surge capabilities and other factors. Nevertheless, the straight line probably is not a bad approximation for average outcomes.

What can be done to change the graph?

Several different schemes for using sorties are possible. First, no sorties could be flown until day two or three. Then, the graph would look the same, but it would start two days to the right of the ordinate. Ignoring possible destruction of aircraft on the ground, this would mean that the full weight of close air support would hit the enemy on his second or third day of operations. Under some circumstances, the sudden onslaught of previously unseen close air support might have as much of an impact on enemy operations as commitment of a large ground reserve. The idea of holding ground forces in reserve for a period of time is well accepted. Should air forces be treated differently?

Second, the line could be kept horizontal by deciding on some level of sorties that could be maintained over time. Although that level clearly would be much less than maximum surge capability, it would be higher at the end of the period than if a standard approach were used.

The last theoretically possible variation is an ascending line, on which sorties on day one start out at a very low level and increase over time. Attaining the same number of sorties on the last day as could be achieved on the first day with a maximum effort is not possible, for obvious reasons. However, flying more sorties on the last day than would be available using the other schemes is possible.

One benefit from varying the sortie pattern comes from the prediction that not every day of battle is equally important for oneself or for the enemy. In fact, effort in war comes in spurts and surges, rather than some inexorable pressure like a flowing river. Lulls between enemy offensive or defensive surges offer opportunities that can be exploited if force is available to do so.

The theater commander would like to be able to concentrate ground and air power to take advantage of these opportunities, but he can't if close air has been expended in some mechanical way. Thus, a sortie may be more valuable on one day than on another. And a sortie that is available on a later day because an aircraft was not previously lost in combat or is not down for needed maintenance may not be a wasted sortie by any means.

Indeed, a sortie saved is worth more than a sortie rashly used. The computer game tended to show that a concentration of sorties, made available by holding back in the beginning of the battle, could be beneficial.

So far, we have discussed producing reserves by rearranging sortie production patterns. The same obviously can be done by holding units out of the battle, as the British did in 1940. One counter argument is that air, because of its mobility, can be shifted quickly from one chore to another and thus constitutes its own reserve. In theory, that may be true. But in practice, at least when the situation has been tense, no one has been willing to relinquish the support he was receiving from air. As we earlier mentioned, the Germans on the Russian front had devoted most of their air effort to close support. They finally realized that their air support should be striking behind the lines, and actually shifted it to do so. The cries of anguish from ground commanders were so loud, however, that the new mission was quickly aborted.[14]

The commander with numerical superiority has a better chance of shifting effort than the commander who must strain to do the minimum things that need doing. The beauty of an air reserve, controlled by the air component and theater commanders, is that it can be thrown in without taking anything away from anyone. Lastly, great advantage can be gained, as the British discovered, if some system allows the rotation of battle-weary units off the front to allow them to rest and recuperate.

Historical examples are too few and far removed in time to establish the absolute need for air reserves. With the theory added, however, the case seems at least strong enough to merit a commander's consideration, especially if he expects to meet a numerically superior foe.

The Orchestration of War

The theater commander and his component commanders are responsible for conducting military operations that will lead to attainment of political objectives specified by the leaders of their country. To do so, they employ the air, sea, and ground forces needed to attain a military objective that supports the political objective of the war. The political and military objectives of both sides together establish the nature of the conflict.

Our purpose here is not to delve deeply into philosophical and theoretical questions surrounding political and military objectives; however, commanders at the operational level must consider the relationships between the two if they are to do their jobs properly.

The political objective of a war can range from demanding unconditional surrender to asking the opponent to grant favorable terms for an armistice. The military objective that will produce the desired behavior on the part of the enemy will be related to the political objective and will in turn heavily influence the campaign plan designed to attain it. Examples from history abound.

In World War II, the political objective of the Western Allies was the unconditional surrender of Germany and Japan. The military objective given to all theater commanders was destruction of the enemy's armed forces preparatory to invasion of the homelands. The Allies actually destroyed most of Germany's armed forces, and most of her war industry, as they drove to-

A B-17 *Flying Fortress* of the 8th Air Force unloads a lethal load of destruction in a daylight attack on Berlin on 29 April 1944.

ward the heartland. In the case of Japan, the Allies strangled the home islands with sea and air power, eventually substituting aerial bombardment for ground invasion.

Both countries surrendered unconditionally because they had lost the ability to protect their people and maintain the integrity of the state. It is not conceivable that either would have surrendered unconditionally while they possessed effective means of resistance. The military objective thus had to be the virtual disarming of the enemy state by destruction of one or all of the branches of its armed forces, or by vitiation of its armed forces by denying them materiel needed to operate offensively or defensively.

The defeat of Germany and Japan exemplify the pure, un-constrained strategic offensive in pursuit of an unconstrained po-litical objective. At the other end of the scale was North Viet-nam's war against the United States. The North Vietnamese had a simple political objective: withdrawal of the United States from Indochina. They estimated that the Americans would only invest a certain amount of time, money, and lives in the enter-prise. If the North Vietnamese could force their enemy to go beyond those limits, victory would be theirs. Thus, the North Vietnamese military objective was to inflict human and mone-tary costs on the Americans over an extended period of time. Given the enormous military power of the United States, the accompanying campaign plan had to be a strategic defensive. It succeeded.

Looking at World War II and the Vietnamese War together allows us to note simple principles that can be quite valuable.

• First, nobody gives up *everything* until further resistance becomes obviously either futile or physically impossible.

• Second, the degree of pain that a state will endure is related to what it is asked to give up. The United States was only asked to give up its position in Indochina—something it did not view as particularly important. Contrast that with the US reaction when Japan asked the United States to give up dominance in the whole Pacific Basin.

• Third, beauty is in the eye of the beholder. Indochina was a sparkling jewel worth any sacrifice for the North Vietnamese. But for the Americans it was a dank jungle with only theoretical interest stemming from containment doctrine.

• Fourth, and related to the last principle, is that the intensity of the fight is established by the side that has the greatest inter-est and will—not by the side with the least. In Vietnam, there-fore, America's campaign could not be based on her level of in-terest in the region if it were to succeed. Instead, it had to be based on what would make the North Vietnamese lose interest.

• Fifth, and perhaps most important, military objectives and campaign plans must be tied to political objectives *as seen through the enemy's eyes, not one's own.* Failure to follow this cardinal precept has led state after state down the primrose path to embarrassment, or defeat.

In many cases, the theater commander may not have much say in establishing political objectives or even military objectives. He may be given both and told to develop a campaign plan simply to attain the latter. In some cases, however, he will be in a unique position to spot logical inconsistencies or physical problems not known or missed at the strategic level of command. The Germans encountered problems of this sort on the Russian front during World War II. In the first instance (spotting logical inconsistencies), the General Staff opposed Hitler's plan to conquer the Caucasus region for economic exploitation before defeating Soviet ground armies.[1] In the second case (spotting physical problems), the political leadership wanted to capture Soviet industry intact; therefore, it forbade the *Luftwaffe* to conduct a sustained campaign against Soviet armament factories in the first part of the war, even though doing so was a compelling military necessity.[2]

In both cases, the supreme command refused the counsel of the military and consequently made the jobs of operational commanders more difficult or even impossible. These disagreements will arise in virtually every war. The operational commander must make his views known, but he also must be ready with contingency plans in the event he is overruled.

Under some circumstances, the theater commander and his component commanders may be able to influence or even write the military objective for the war. What must they consider? The military objective must lead to the political objective, and the commander must have in place, or have guaranteed to him, adequate forces to attain the objective. Military objectives will vary substantially, but they tend to fall in one of three categories.

First, the military objective can be the destruction—or neutralization through maneuver—of some or all of the enemy's forces. The necessary degree of destruction will depend on the importance of the political objective as seen by the enemy. It also will depend on the enemy's capacity, as was the case in the Falklands war. Great Britain won by destroying enough of the Argentine air force and navy that further Argentine operations on the Falkland Islands were impossible. The Argentineans were beaten even though large ground forces were available on the mainland.

Second, the military objective can be the destruction of some or all of the enemy's economy, especially his war-related economy. Japan surrendered because she no longer was able to protect herself from aerial attack and her industry was collapsing around her. Similarly, the destruction by air bombardment of petroleum production and internal transportation networks made Germany unable to wage modern war by the end of 1944. In both cases, destruction of key war industries was either complete or threatened to become so very quickly. Similarly, but at the other end of the scale, Israel attained her political objective of stopping Iraqi nuclear arms development by using her air force on 7 June 1981 to destroy Iraq's nuclear research center— the military objective.

Third, the military objective in support of a political objective, such as unconditional surrender, can be the destruction of the will to resist, either the will of the government or the will of the people. This objective is tenuous because it is difficult to get at "will" without destroying either armed forces or economy. In other words, the will to resist collapses when the armed forces no longer can do their job or when the economy no longer can provide essential military—or civilian—services. Judging the will to resist also is very difficult, as the following example indicates. Almost anyone in 1941 would have judged the Germans to be more rational than the Japanese, yet it was the Japanese who surrendered when reason demanded it—even while significant ground and air forces were still intact.

While direct attacks on the enemy civilian population may seem a viable way of breaking the national will (moral objections aside), difficulties abound. Populations have proven resilient, and the people may have little influence on the decisionmaking of their rulers. The indirect approach to the people is another matter. The North Vietnamese essentially attacked the will of the American people to resist. Similarly, the Allies in World War I succeeded in significant part because the economic blockade made life in Germany almost unbearable.[3] In World War I, however, the Germans were not ruled by an irrational tyranny.

One other possibility should be considered. The will to resist sometimes can be broken with kindness rather than with destruction. In retrospect, it seems possible that the Germans could have defeated the Soviets had they gone into the Ukraine and other colonized areas as liberators, rather than as destroyers.

When a military objective has been chosen, it is necessary to act. To help guide the thinking needed to decide on a course of action, an analogy from personal combat is useful. When two antagonists meet, they have several choices. They can stand toe to toe and trade blows with each other until the weaker antagonist finally collapses. Assuming roughly equal strength at the outset, the outcome largely will depend on who can absorb the most punishment.

Another possibility is for one fighter to move in, strike a blow, and retire before the opponent can respond. In this case, the winner probably will be the smarter and faster fighter. If he has executed well, he may not suffer much in the process.

The third approach involves doing something unexpected, like using a club, pulling a third party into the fight, or striking indirectly at the antagonist by kicking over his water bucket or eliminating his manager. Given enough time, any such stratagem may do the trick at little cost.

The American bomber offensive against Germany in World War II started out like two fighters exchanging blows. It almost failed because the Germans were able to knock down too many machines and crews. The offensive shifted gears with the introduction of long-range fighter escort. The escorting fighters then took over the role of the fast-moving, in-and-out puncher and kept the Germans sufficiently at bay to permit the bombers to do their work. Finally, the bomber offensive turned on German petroleum and thereby removed the sustenance of the entire war machine.

Could the order of the offensive phases have been reversed? In retrospect, early destruction of the petroleum system would have shortened the war and would have been a quicker means to an end than was the campaign that started out directly against aircraft and aircraft manufacturing.

The analogy from personal combat intimates that the first possibility, the toe-to-toe slugging match, is the least attractive and the least likely to be adopted. In fact, it is the orthodox approach and the one that most commanders have adopted for most campaigns and battles recorded in military history. The choice of an in-and-out campaign or of an indirect approach has been relatively rare and therefore is, by definition, radical. The orthodox approach also is, at the risk of oversimplification, the typical American approach and has been so since General Ulysses S. Grant's successful siege of Vicksburg[4] during the US Civil War. It certainly was the high command choice in World War II, when Marshall envisioned a relentless frontal attack on the German homeland, powered by American industry and led by overwhelming firepower.[5]

The initial concept of the air campaign against Germany in World War II was the same—a massive frontal assault that would roll over the opposition. The Southwest Pacific campaign was an anomaly in the last century of American military history, as were the first six months of the Korean campaign.

This fact of history, and especially of American military history, is relevant for the operational commander who might propose something other than orthodoxy. He must realize that officers on his own staff, and especially in higher staffs, will strongly oppose his "radical" ideas. They will do so with the best of motives, sincere in their belief that they are protecting against flights of fancy and against reckless adventures that may well lead to disaster. *They could be right.* The burden of convincing the deciding authorities that they are not will rest with the operational commander.

One could argue that the orthodox approach has worked acceptably well for the United States for more than a century and that it would thus be foolish to change it. Such might be the case if conditions remain similar to those encountered in past wars. In every war the United States has fought since the War of 1812, the country has had an industrial base capable of outproducing by far her every enemy. In the twentieth century, she also commanded the combined technology-production front. In any war to be fought in the future, the Americans may be able to follow the old conservative ways, as long as the enemy is disadvantaged in materiel and technology. If the enemy is not, then the old ways may be a recipe for disaster. Convincing key subordinate and superior officers that conditions have changed sufficiently to invalidate more than a century of experience may be exceedingly difficult.

The conceptual problems just discussed apply to all of our cases, but they are particularly pertinent to Case I, in which both sides in the conflict commence with approximately equal vulnerabilities. If the enemy is equal or superior in manpower and production, a frontal assault may not be the answer. If he is decidedly inferior, then it may not matter—although the conservative approach may take longer and cost more in blood and treasure than would an alternative. In chapter 3, we developed outlines for an offensive air superiority campaign and saw illustrations from the American experience against Germany in World War II. The Army Air Forces conducted that campaign very traditionally—even though the objective of the campaign was radical and untried—at least until the time when bomber losses became almost prohibitive.

The same thing possibly could be done in a situation where both sides have vulnerable bases, although actual conduct of the campaign would be made more difficult by enemy action against base areas. What general rules, then, can be adduced for conducting a campaign?

CENTER OF GRAVITY MUST BE IDENTIFIED

In all cases, the enemy center of gravity must be identified and struck. For the air superiority campaign, that center of gravity can be any of the areas enumerated in chapter 3—enemy air equipment (aircraft and missiles), enemy logistics, enemy personnel, or enemy command and control. In choosing the appropriate center of gravity and in devising means to strike at it, the disposition of enemy forces can be especially important.

The *Case I* situation evokes the image of two opposing lines, or perhaps concentric circles, of opposing castles, where either side can sally forth and even besiege an individual castle—mindful, of course, that help for the besieged will shortly come from elsewhere in the line if action is not taken to block it.

The *Case II* situation suggests besiegers surrounding a castle and attacking it at their leisure. The *Case III* situation brings to mind a castle surrounded by besiegers, under constant bombardment, and able only to defend. Finally, *Case IV* could be likened to a battle taking place midway between two castles, where the occupants of both castles could hurl missiles at the combatants, but not at the other castle.

Extending the castle analogy, we can conceive three basic approaches to defeating the enemy.

The *first* is a broad front approach where the object is to reduce every castle, either one by one or, if sufficient forces are available, by simultaneous attack.

The *second* approach is to reduce one or two castles, ignore the remainder, plunge through the gap, and win by seizing the capital.

The *third* approach is to figure out a way to avoid the castles entirely and go directly to the political center of gravity—the capital or the king.

Assuming that all three approaches are physically practical, the third promises to be the quickest and cheapest, the second the next best, and the first the slowest and most costly.

The war against Japan could have been, and actually was, prosecuted using every one of these approaches. But the second and third proved decisive and provide excellent models of how to plan an integrated campaign that uses all available resources.

In the spring of 1942, the perimeter of the Japanese Empire ran from the Aleutians to Wake, through the Solomons to New Guinea, from New Guinea to Singapore via the Netherlands

East Indies, and thence back to Japan by way of Burma and China. The Japanese had roughly a 12,000-mile front that in its most crucial areas was 2,500 miles from Tokyo. Following the Battle of Midway, the Japanese found themselves unable to move closer to Hawaii or to progress in the Aleutians. Consequently, they intended to continue operations in China and Burma while making two major moves against the Americans. The first of these major moves was in the southern Solomons, where a base on Guadalcanal would give them the ability to interdict American shipping to Australia. The second was a move across the Owen Stanley mountains to Port Moresby on the south coast of New Guinea, from where they could threaten Australia directly.[6]

What were the Allied choices to counter the Japanese?

One must remember that before World War II had started for the United States, President Roosevelt and Prime Minister Churchill had agreed that the first priority would be the defeat of Germany. Until victory was assured in Europe, the Allies would stay on the strategic defensive in the Pacific. The emphasis on defeating Germany first remained in effect throughout the war. But a combination of political and military factors quickly transformed the strategic defensive in the Pacific to an expedient offensive, albeit an offensive that suffered because the European theater had first call on resources.

In a certain sense, the strategy against Japan at the highest level was a broad front approach, with counterattacks to be launched in Burma, China, the southwest Pacific, the south Pacific, the central Pacific, the north Pacific, and finally, completing the circle, in the northwest, when the Russians could be induced to enter the conflict. In fact, resources simply were not available for large-scale operations outside the southwest and central Pacific areas, so the grand broad front approach never was seriously executed.[7]

Any of the three approaches to reducing the Japanese citadel could have been chosen, even in the relatively bounded area of the southwest through central Pacific. Keep in mind that this "bounded" area created a front of almost 6,000 miles. As a help in visualizing the enormity of the area, consider that New

A B-24 *Liberator*, with landing flaps down, skims over nesting goonie birds on Midway Island 21–24 December 1942 during a 7th Air Force bombing mission to Wake Island.

Guinea alone was longer than the entire Russian front, from the Caspian Sea to Leningrad. Early in the war, the idea was to move forces toward Japan for an eventual invasion of the homelands. Given the geography, these forces necessarily would have to move from island to island, and all would have to be reduced in the process. Air was seen as an adjunct of ground and naval surface forces.[8]

The Joint Chiefs of Staff, in July 1942, directed that this methodical reduction process be started in the Southwest Pacific theater (under General MacArthur) and in the South Pacific theater (under Admiral Robert Lee Ghormley, commander of Allied naval forces in the South Pacific in 1942, who directed the attack against the Solomon Islands in August 1942). Admiral Ghormley was to move up the Solomons chain toward the great Japanese base at Rabaul on New Britain, while General MacArthur was to take northeast New Guinea, finish off the Solomons, and finally lay siege to Rabaul.[9] After this part of the Japanese line was destroyed, the Joint Chiefs envisioned peripheral movements to reduce Japanese positions in the rest of New Guinea to the west and in the Ellice, Gilbert, Marshall, and Marianas islands to the east and north.[10]

In this early period of the war, the principal commanders, including General MacArthur, continued to see air as a supporting arm. The evolution in thinking, especially on MacArthur's part, makes the Pacific campaigns so important to study and understand.

Recall from chapter 2 that MacArthur had made the radical decision to conduct an intermediate campaign for air superiority. He was still thinking conventionally, however, in that he believed it necessary to reduce every Japanese redoubt—or "castle," to continue the earlier analogy—in his area. He still thought it necessary to occupy Japanese bases in northeast New Guinea, then those in the Bismarck archipelago, and finally to seize Rabaul at the north end of New Britain. After carrying out this conventional campaign, based on an unconventional concept, he intended to continue the reduction of Japanese positions on the rest of New Guinea, and then move further west toward Borneo before starting island by island towards the Philippines.

Then a radical idea came from an unexpected source, the Joint Chiefs in Washington.[11] By the late summer of 1943, MacArthur had made significant progress in the Huon Peninsula area of New Guinea (Lae, Finschafen, and Salamuau). This campaign was to lay the base for the eventual investment of Rabaul. The Joint Chiefs, however, in conjunction with the British at the Quadrant meeting in Quebec in August 1943, directed MacArthur to complete the Huon Peninsula operation and then to move along the New Guinea coast toward Volgelkop. He was directed to bypass Rabaul, to leave it for neutralization by air attack as required. MacArthur had opposed leaving Rabaul unoccupied, because in conventional terms it was a serious threat to his right flank.[12]

Although MacArthur had strongly opposed bypassing Rabaul, the Joint Chiefs directive to do so started him thinking about the possibility of bypassing other places. Combining the concept of an air superiority campaign with the concept of by-passing enemy positions, MacArthur envisioned a campaign that would take him to the Philippines, while whole armies of Japanese—but not Japanese air armadas—remained intact behind his lines.

In brief, General MacArthur captured only those areas necessary to support air operations against Japanese airfields, and then used the captured fields to extend air superiority out as far as possible. Air superiority established, he jumped over intervening Japanese ground positions to occupy new bases from which air superiority could be further extended. He continued this process to Leyte Gulf, where he leaped beyond the range of his land air, and depended entirely on the cover of carrier air. His decision might have led to catastrophe, because portions of the Japanese fleet succeeded in penetrating Leyte Gulf after the US 3rd Fleet moved away in pursuit of another portion of the Japanese fleet.[13]

General MacArthur recognized his error and backtracked to establish land air bases on Mindoro—which he had intended to bypass—to support his invasion of Luzon.[14]

The foregoing has been a necessarily brief account of concepts and operations that carried MacArthur to the Philippines in January 1945. The key lessons are several:

• Air superiority can be an end in itself, at least on an intermediate basis.
• Ground forces (and naval forces) can serve as an adjunct to air forces in the battle for air superiority.
• Penetrations on a huge operational scale can be made—MacArthur's was over 2,000 miles deep and very narrow—if the flanks can be covered by air, or sea.
• And bypassing or ignoring pockets of great strength can be feasible when they are neutralized or isolated.

Within the context of these very broad lessons, the Southwest Pacific theater, as we saw in chapter 2, also offers rare insights in achieving air superiority. Additionally, the war against Japan illustrates the possibility of ignoring completely the line of "castles" and going directly to the political center of gravity.

USAF Photographic Collection, National Air & Space Museum, Smithsonian Institution

A B-25 *Mitchell* of the 5th Air Force takes part in a bombing run over Boram Field, Wewak, during the New Guinea campaign.

Identifying air superiority as the objective in the Southwest Pacific theater did not make the gaining of it happen. What made it happen was the application of the general principles already discussed. In the Pacific, the means did not exist in the first three years of the war (up to 1944) to do much about the sources of aircraft and pilots. Similarly, there was no apparent way to attack enemy command and control on a theater basis. That lack led to direct attacks on aircraft and personnel, attacks on logistics, and exploitation of doctrinal weakness. We saw in chapter 2 how General Kenney, in a series of well-conceived operations, attacked and exploited all three areas.

The theater commander must decide whether to adopt a broad front or indirect approach—or something in the middle. Ideally, he would make a decision even before the war began that would provide a coherent plan leading to victory, even if that victory were years away. Given the uncertainties, the friction and fog of war, the fact that he faces a human enemy who will attempt to impose his will during the course of the conflict, and perhaps most important, the fact that he will grow and learn as the days pass, it is unlikely that he will be able to craft a perfect plan. Thus, he must be flexible and ready to change when circumstances demand it.

Having chosen or been assigned a military objective, and having chosen at least his initial broad plan, the theater commander must determine whether he can best attain his objective with air, sea, or land forces. All three probably will not have an equal role. If they do not, the key force must be identified, so that the other two can stand in support.

The concept of identifying a key force may make some people uncomfortable, especially if they are in an environment where inter-Service cooperation is stressed, and where cooperation means either subordination (air support of the army, for example) or that each Service has a precisely equal role, regardless of circumstances. In former times, everyone accepted the fact that some things could be done only by a navy, and some things could only be done by an army. On occasion, a reason might exist for them to work together. But for the most part, each had a specific job to do that if done correctly would lead to meeting the objectives of the war. The only difference between then and now is the third dimension of air power—which must be evaluated as a potential key force in the same way that the army or navy in the past was evaluated.

To understand the concept of a key force and the relationship · among complementing forces, thinking about another art form—the concerto—is helpful. A composer writes a concerto to say something, to attain some objective. Having selected an objective, the composer decides how best to reach that objective. Should it be a piano concerto, a violin concerto, or a flute concerto? Only one will get him to the objective he has chosen; clearly, a piano cannot say what a violin can say, and vice versa. That he has chosen an instrument to be his key force does not mean that the other instruments do not have roles. To the contrary, the other instruments are vital, for they provide the support that allows the key force to do things it could not do by itself.

During the course of the concerto, the key force will be the only instrument active at certain times; the rest are in repose, awaiting their turn. At other times, the key force is silent while the complementing forces bear the whole burden. The composer, and later, the director, has the task of orchestrating—not subordinating nor integrating—his instruments so that each can do its job—whether that be as the key force or the supporting force. In the process, he does not try to make one instrument sound like another, or do another's job; rather, he uses each to do what it is naturally constituted to do and what only it is capable of doing.

Orchestration, not subordination or integration, is the sine qua non of modern warfare.

If we carry the concerto analogy to the realms of warfare, we can say that a particular war, or campaign, or phase of a campaign could be a sea concerto if sea power were the key force. Likewise, we would say that a war or campaign was an air concerto if air forces had the dominant role. We also would say that the theater commander had the job of *orchestrating* his forces in such a way as to achieve his objectives. How does the commander decide what his key force in a campaign should be?

The easiest of the three forces to choose or reject is the sea. It is clearly not appropriate if the campaign is against a continental power that has little sea commerce and where the area of hostilities is not bordered by oceans. On the other hand, it may be entirely appropriate if the campaign is against an island power which can be isolated and starved into submission if its sea lines are cut. If the sea is chosen as the key force for the campaign, air may still be crucial to allow appropriate sea operations. Ground forces may be needed to take or occupy land areas controlling critical sea passages.

Choosing between ground and air as the key force for the campaign is far more difficult; given enough time, money, and blood, either can theoretically accomplish what the other can do. That is, to kill every enemy ground soldier by air attack is theoretically possible, and capturing and controlling enemy means of production with ground forces also is obviously possible. Let us begin the decision process by identifying clear-cut cases.

Ground must be the key force if air cannot make a substantial or timely contribution to the campaign effort. Air is of marginal value in a fight against self-sustaining guerrillas who merge with the population. In this case, no significant target exists for air attack. Ground also should be the key force if short-term occupation of limited pieces of territory is the military objective for either side and will in itself end the war—as did German occupation of Prague in 1939. In the very short term, air cannot stop large bodies of men; interdiction takes time to work; and attacks on war production take even more time. Ground must be the key force if time is of the essence, and it is agreed that ground action can lead to the political objective significantly faster than could air action.

To some extent, time drove British and American strategy against Germany. Quite conceivably, Germany could have been defeated through air attack. But the certainty of her defeat by air alone decreased if the Soviet Union made a separate peace with the Germans. Veiled Soviet threats to do so, if a second front in France was not opened expeditiously, helped drive the British and Americans to choose the ground approach, not the air approach. Before discussing air as a key force, one must remember two points about ground campaigns.

First, territory is a dangerous enchantress in war. Serious wars are rarely won by capturing territory, unless that territory includes a vital political or economic center of gravity, the loss of which precludes continuing the war. The capture of France in 1940, significant though it was, did not win the war for the Germans. France was not the center of gravity of the antiAxis coalition—even before the United States entered the conflict. After World War II, the United States, not Western Europe, became the center of gravity in any conflict between the Soviets and the western powers. Territory may well be the political objective of a campaign, but it rarely should be the military objective. Territory will be disposed of at the peace conference as a function of the political, military, and economic situation at the war's end.

Second, assumptions about time are apt to be dangerous. Few things are more difficult to predict than how long a war or a campaign will last. Germany planned for a short war and was unable to endure a long one. Outside observers were almost unanimous in predicting that the Soviets would fall by Christmas 1941. MacArthur talked about sending troops home for Christmas from Korea in 1950. Johnson's "end of the tunnel" prediction was tragically wrong in Vietnam. In contrast, British and American forces covered more ground after the breakout from Normandy in three months than they had planned to cover in a year.[15] Territory is beguiling and time deceiving: The commander must beware of both.

Air must be the key force when ground or sea forces are incapable of doing the job because of insufficient numbers or inability to reach the enemy military center of gravity. The German campaign against Britain after Dunkirk was based on air, because the army and navy could not get at the British. Although a submarine campaign was underway against Britain at the same time the Battle of Britain was in progress, the submarines could not defeat Britian, nor could they establish the conditions needed for invasion. The German navy, then, was—or should have been—in support of the primary air campaign.

Air may be the key force when enemy ground forces can be isolated or delayed while air works directly against political or economic centers. Similarly, air could be the key if enemy power were confined to a relatively small area, such as an island. Pantelleria, an island between Malta and Tunisia, surrendered after intensive air attack,[16] and Malta, as previously mentioned, was on the verge of surrendering. Air may be the key for a phase of a campaign that is leading to a point where sea or land becomes dominant. It should be the key if the military objective of the war is destruction of the enemy's war production capability. Lastly, air may be appropriate to select as key, under an even wider variety of circumstances if time is not a significant constraint.

In the last several paragraphs, we have suggested guidelines for deciding what the key force of a campaign should be. Making the decision will frequently be difficult, but it is a task that cannot be shirked. Once decided, each participating component can see what its role is. When all these things are known, the jealousy and suspicion that are part and parcel of such an intense

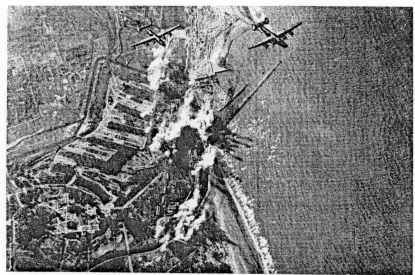

USAF Photographic Collection, National Air & Space Museum, Smithsonian Institution

B-24 *Liberators* of the 2nd Bomb Division, 8th Air Force, bomb enemy harbor installations at Dunkirk, France, on 15 February 1943.

human activity as war will be less likely. Just as a concerto must have a key force to meet the objective of its composer, so must a war plan.

Unfortunately, however, there have been many wars where the "composer" had only an amorphous objective and failed to identify a key force, and where each "instrument" either thought it was dominant or didn't realize what its role was in producing a coherent performance. Wars of this kind have been expensive—and frequently fatal.

The concepts of objectives and orchestration presented, we can now turn to the air campaign itself, whether it is intended as the dominant or supporting instrument of the concerto of violence.

10

Planning the
Air Campaign

The air campaign may be the primary or supporting effort in a theater. In either event, an air campaign plan is a necessity. The plan should describe air centers of gravity, phasing of operations, and resources required. It must provide general guidelines for division of effort among air superiority, interdiction, and close air support. It should explain how other arms will support or be supported.

Like the overall theater plan, it must carry through to the conclusion of the war. We will address some of these areas in detail, but first let us review some critical concepts.

The nature of the enemy is quite important, especially if the air campaign plan envisions anything other than straight attrition. More ways exist to categorize an enemy than can reasonably be integrated and used. For example, he may be rational, irrational, fanatic, rigid, flexible, independent, innovative, determined, or doctrinaire. To the extent that an enemy can be assigned to any of these categories, his plans may be anticipated, and the way he will react to a new situation can be predicted. History will provide some help in assessing the enemy, although it would be foolish to suppose that straight line projections of past behavior are going to be absolutely valid.

The other side of knowing the enemy is knowing oneself.

Making plans requiring a high degree of initiative and independence at every subordinate level may be risky, if peacetime training has emphasized detailed operations orders issued from high staffs. Likewise, if one's military has been thoroughly grounded in attrition warfare and direct attack, selling a maneuver or indirect attack campaign plan may be quite difficult. This is not to say that such plans should not be proposed; they should, of course, but the author must be prepared for strenuous, honest opposition from above and below. The most important part of assessing one's own side is honesty. New tactics can be learned during the course of a war, but it is unlikely that the same applies to new modes of behavior. Therefore, the commander must accept the fact that he has certain human material with which to work and that everything must be built around the reality of his forces, not on how he would like them to be.

Although preceding chapters have gone into great detail on air superiority, interdiction, and close air support, the commander must decide in the campaign plan how these three elements of air operations are going to be integrated. The right decision can lead to victory, the wrong decision to defeat. Let us start this difficult process by considering air superiority.

We maintain that it is not possible to win a war if the enemy has air superiority. Indeed, no nation enjoying air superiority has ever lost a war by the force of enemy arms. A commander who tries to win—or not lose—without air superiority is trying to do what no one has done before. Thus, the prudent commander will do what is necessary to become superior in the air. (Conceivably, he could even withdraw on the ground to create conditions more favorable for winning air superiority.) The first thing he ought to do is make an assessment of where he stands with respect to his enemy.

Recall from chapter 1 the five cases that can confront the commander as he starts his air superiority plan or campaign. *Case III*, where the enemy's bases cannot be attacked, only allows for defense against enemy onslaughts. The situation is somewhat similar in *Case IV* where air operations can take place over the front, but not in either contender's rear area. In this case, the commander must decide how he is going to meet the enemy over the lines. *Case V,* with no combat air, is the simplest; only contingency plans against introduction of air need be made.

Two cases remain, and they are the toughest because they offer the most possibilities for good or ill. Of these, *Case II* is the most benign as it permits action against the enemy's bases while one's own bases are essentially immune to attack. In *Case II,* the biggest problem facing the commander is what elements of the enemy air system to strike in order to win air superiority. The last case, *Case I,* is by far the toughest, for it has no sanctuaries: What one side can do, the other side also can do. Case I is a chess game, and like a chess game, the statistics favor the side that moves first.

Some observers will maintain that a plan calling for theater air superiority is too ambitious and that it proposes more than is necessary. In its stead, some will recommend local air superiority. "Local air superiority" can have two meanings. Most often, it means establishing cover for a surface operation. In its other sense, it suggests a phase in an air campaign similar to a breakthrough operation on the ground that establishes the base for further similar moves. This second definition merely recognizes the reality that achieving theater air superiority is rarely possible in one battle.[1]

The first definition proposed local air superiority as an adjunct to a surface campaign. Unless the time needed for local air superiority is very short, it is an unsound concept because it throws air onto the defensive. We have seen rather clearly that the defense is at a distinct disadvantage in air war. Let us examine two illustrative situations.

First, local air superiority could make good sense for a short operation, such as passing a naval fleet through a strait or near a land mass. Attacks can be made on enemy air bases to hinder flight operations, and enough cover can be put over the fleet to protect it while the enemy is trying to organize large scale attacks. Total time required for local air superiority should be on the order of hours and is achievable.

In the second case, the proposal might be to provide local air superiority for a ground counteroffensive or defense. Here, the time required moves from hours to days at a minimum and perhaps even weeks or months. Now the enemy has ample time to concentrate forces against the covering air and can take full advantage of the air attacker's significant ascendancy over the air defender. We have seen example after example of the force ratios needed to defend and the difficulties inherent in reacting rather than initiating.

A kind of compromise exists between local and theater air superiority advocates that favors the theater side. If the overall theater campaign plan envisions winning the war by occupying a piece of territory, as opposed to destroying enemy forces or military production, establishing air superiority over the disputed territory may be accomplished by driving enemy air back to the point where it cannot reach the battle zone. This concept is entirely different from the concept of a covering operation. Such a campaign plan subjects the enemy to attack, while protecting one's own bases (a *Case II* situation).

In the case where both sides are fully vulnerable to attack (*Case I*), the commander has the option to operate offensively or defensively. Although he might have reasons to go on the defensive initially, the air commander must be predisposed toward the offensive. He should go on the offense unless he finds compelling reasons for not doing so. If he chooses the offense, he then must decide on targeting priorities.

CENTER OF GRAVITY MAY NOT BE REACHABLE

Targeting priorities will be a function of perceived enemy air centers of gravity. We covered in chapter 3 possible centers of gravity in some detail. We must keep in mind, however, that the real center of gravity may not be reachable initially. Defensive considerations may compel the commander to strike first at something other than the final objective. Suppose the enemy has a dozen airfields that are especially well suited for offensive operations. These fields may not be important in the long term, but could support damaging enemy strikes in the short term. These fields, then, might be the first order of priority, the first phase in the campaign. Likewise, neutralizing a portion of the enemy's ground-based defenses may be necessary for the campaign to develop as planned. In other words, the route to the center of gravity may not be a straight line.

The air superiority campaign (whether an end in itself or a means to an end) should not be waged with air assets alone. Naval and ground forces should play a role wherever possible. The more innovative their actions, the more likely are they and the campaign to succeed. We already have seen how the British sent commandos to knock out an effective German bomber unit on Crete, how MacArthur and Kenney used ground forces to seize airfields, and how the Israelis from the 1973 war to the 1982 Lebanon incursions used naval and ground forces to knock holes in ground-based air defense systems. The Israelis even won complete air superiority without use of air weapons at Entebbe.[2] On that operation, a group of commandos by themselves destroyed the enemy's air force. If theater and component commanders are intrepid and innovative, and if they understand the overriding need for air superiority, they will work together to win it.

In the process of planning or executing an air campaign, three especially thorny issues confront the commander and his planners.

- The first is the use of air in emergency situations, such as a fast-progressing enemy ground offensive.
- The second is deciding on the relative effort to be assigned interdiction and close air support.
- The third is the desirability of carrying out air superiority, interdiction, and close support simultaneously.

Few things are more disconcerting than a sudden, massive enemy offensive that is either progressing well or seems on the verge of doing so. The tendency is to throw everything against the ground movement and to stop air superiority and interdiction operations until the emergency is over. This tendency, although natural, may be deadly—especially if the enemy's air force is still capable of fighting effectively. When one throws everything against the leading edge of a ground offensive, pressure on enemy air decreases significantly, and perhaps to the point where the enemy can undertake previously impossible counterair operations. If everything is concentrated on a ground objective, the enemy can concentrate his air offensively against the aircraft working in support of ground forces. Or he can take the opportunity to press his own air superiority operations aggressively. In either event, the enemy will realize the advantages accruing to the offense.

Given all these problems, the commander may be correct in throwing everything at the ground under the following circumstances: If the battle in progress is unquestionably the decisive battle of the war; if withdrawal is militarily impossible; if losing the battle means surrender; if the battle certainly will end within a few days; and if stopping the enemy positively means no further enemy offensive before friendly air and ground forces can be rebuilt.

If all of the above conditions cannot be met, diverting every effort to ground support makes subsequent success problematical, even if the immediate threat is stopped. The prescription is clear, but no one will want to take the medicine.

The *second thorny issue* is the allocation of air between interdiction and close support. Only for those countries in which one or the other is doctrinally anathema will the decision be easy. Previously noted were the Israeli distaste for close air support and the Soviet full embrace of it in World War II. Where the problem is not doctrinally solved before the fact, commanders and planners must wrestle with it. The easiest way to start is by asking if either is clearly inappropriate. If nothing is at hand to interdict, such as in a low-level guerrilla war, then all can be given to close support. Unfortunately, no immediately obvious example comes to mind of where close support would be pointless. Thus, other approaches must be used to arrive at an answer.

With few exceptions, sorties flown in close support will reduce the sorties that can be flown in interdiction, and vice versa. One exception obtains when an air force has aircraft specifically designed for close support that cannot survive in an interdiction environment, at least until air superiority is won. If other things are equal, these planes might as well be used for close air support—if they don't require protection by other aircraft that could be better used elsewhere.

The interrelationship between close air support and interdiction sorties demands that the commander, the theater commander in this case, decide which one will most benefit his plan. The weight of history, as well as logic, falls on the interdiction side. We covered the history of both in chapters 6 and 7. Materiel and troops are easier to keep away from the battle than to engage at the front. They are easier to destroy when they are in assembly or configured for movement than when they are deployed to do battle. Carrying the thought to the ultimate, one pictures one bomb on one tank factory potentially causing scores or hundreds of tanks not to be built. Conversely, the best that one bomb can do at the front is knock out one tank that already may have paid for itself in damage done.

If the primary emphasis (after air superiority) is going to be on interdiction, the interdiction can be either distant, intermediate, or close, as described in chapter 6. The *distant* is directed at the source of enemy supply; the *intermediate* at bivouacs, transportation nodes, depots, and theater movement targets; and the *close* at movement very near the battlefield. The degree of tactical coordination with the ground component commander is very high for close operations, less for intermediate, and conceptual for distant. Close interdiction should have a major effect on the ground battle; thus, the air commander must direct operations that meet the ground commander's explicit needs. This does not mean that air should be subordinated to the ground commander.

The *third thorny problem* confronting the commander is the likelihood that he will be asked to conduct two or three missions simultaneously. For example, an adverse ground situation may lead to requests for close air support and close interdiction, while air superiority still hangs in the balance. We have discussed the theory of this possibility—and the theory provides an easy answer. Unfortunately, the real world frequently refuses to bow to even the best theory, so we can't take complete refuge in that answer. To address this problem, the commander might, in desperation, divide his air forces into three equal parts, devoting one to air superiority, one to interdiction, and one to close air support. Except in the most extraordinary circumstances, this division surely would be the wrong answer. Very few situations would be so symmetrical as to indicate such a division. In fact, the chances would be quite high that not a single one of the "thirds" would be capable of carrying out its mission. Consequently, all might fail disastrously. What is the commander to do?

Concentration probably is the most important principle of air war. Therefore, the air commander should make every effort to convince his ground component commander brother, and his theater commander, that they should all choose some mission which a concentrated application of air power could bring to fruition. In this decision process, the commander must remember how dangerous it is to try other missions before air superiority is won. Also worth emphasizing is the fact that air power has been more useful in interdiction than in close support. (We saw earlier that the German army decided too late on the Russian front that it should have asked the *Luftwaffe* for interdiction rather than close air support.) Given the critical importance of air superiority, and the historical success of interdiction, the possibility exists of proposing a compromise solution to demands that all three missions be carried out simultaneously.

Clearly, air superiority must be the first air priority because so much else—ground operations, close support, and interdiction—is heavily dependent on it. Thus, conceptually, an interdiction effort should not commence before the air superiority campaign is obviously on the road to success—when enemy air is no longer crossing the front and can no longer defend effectively against interdiction operations. As earlier suggested, however, an area exists for logical compromise, an area that will benefit both missions. Systems exist that support both enemy land and air operations. Their precise identity will vary from war to war. But for the foreseeable future, the petroleum net will be a strong candidate, as will the transportation net if it can be hit behind the enemy airfields it is supporting.

Another potential target is the enemy's theater command and control system. Good intelligence and thorough analysis should produce more candidates. To the extent that systems mutually supporting air and ground can be identified and struck, mixing interdiction and air superiority makes good sense.

So far in this chapter, we have looked at how the air commander constructs the air campaign, in terms of air superiority, interdiction, and close support. These three elements are the main elements of air warfare. However, other elements, while less encompassing, cannot be ignored. The remainder of this chapter will deal with the more salient of them.

Confederate General T.J. "Stonewall" Jackson once said that the duty of a commander is to "mystify, mislead, and surprise." His injunction applies as much to the air commander as to the ground commander. Indeed, deception can be a powerful weapon. Few things could be better than to make the enemy face the wrong way or drop all his bombs on a useless piece of desert. Deception might lead the enemy to think that an attack would consist of 10 aircraft, when in fact it had 20.

Unfortunately, creating successful deception plans is not easy. Successful plans take into account the nature of the enemy, what he thinks about his enemy, and one's own modus operandi. Deception is a difficult subject, but a few examples from the past may provide some ideas for the future.

One of the most successful of all deceptions was not meant to be a deception, but worked so well that it might work again. As we have already noted, at the height of the Battle of Britain, the British bomber command made a militarily useless raid on Britain that so infuriated Hitler that he allowed the *Luftwaffe* to turn on London. The turn on London almost certainly was the one German error that most influenced the battle in favor of the British. Supreme commanders shouldn't make decisions based on ego or emotional desire for revenge—but they have for thousands of years. They will continue to do so. If their egos can be attacked, they might do the most welcome things.

On an operational campaign scale, we saw how General George Kenney faked construction of two airfields and literally invited the Japanese to attack them. Meanwhile, he was secretly building the real airfield that would allow his fighters to escort his bombers to Wewak—a field the Japanese "knew" was safe

Major General Claire Lee Chennault (center) personally conducts members of the Chinese Aeronautical Affairs Commission on an inspection tour of his headquarters unit of the 14th Air Force at an advanced air base in China in October 1943.

because it was out of range of fighters. *Anything* the enemy "knows" can't be done is well worth doing.

On a still smaller scale, General Claire Chennault, as head of the American Volunteer Group in China, wanted to make the Japanese think his forces were much larger than the 40 or so fighters that actually existed. To deceive the enemy, he periodically repainted his aircraft so the Japanese would think they came from different units.[3]

The possibilities of deception are endless, and virtually no rules exist as to foul or fair. As Churchill said, "In war time, truth is so precious that she should always be attended by a bodyguard of lies."[4]

Related to deception is psychological warfare. It has been most effective when the enemy nation was made up of peoples forcibly included in it. When such potential fissures exist, they must be exploited with every possible means. Generally, this kind of psychological warfare will be waged at the strategic and grand strategic levels. Nevertheless, the operational air commander can do some things. He should certainly make it known that he welcomes defectors, that he will reward them, and that they may volunteer to join in the fight. He must devise ways to get the invitation to the enemy, and then must devise ways that allow an enemy pilot to surrender himself and his aircraft. Given enough defectors, he can establish a squadron or wing of defectors, hopefully flying under a banner raised at a higher level. Not only will the existence of the defectors encourage more to follow suit, the units are likely to be quite effective. They and their aircraft may even be used to penetrate enemy defenses for special missions.

In chapter 8, we discussed reserves. The commander must decide whether he is going to have them and when he is going to commit them. His assessment of the length of the war is important to the decision. If the war will certainly end in one or two days, or with one very short decisive battle, reserves may not be useful. If the war is going to last beyond a couple of days, then the commander probably should opt to hold reserves for reasons previously enumerated. To illustrate the negative case, we saw how the Israelis commmitted their entire air force (minus eight fighters on home combat air patrol and four on runway alert) in a bid for air superiority on the first day of the 1967 war.[5]

This was an instance where a single battle was decisive; it would have been an error to have reserves or not to commit them if they existed. The 1967 war, however, is the only major war fought in the twentieth century where the whole war was essentially decided by a single battle on a single day.

The decision made to maintain reserves, the commander must then adopt a principle for commitment. We discussed the error of piecemeal commitment: If the commander is going to commit the reserves, he should do it in mass to capitalize on shock and surprise. As to where he commits, he has two choices. He can reinforce his own success, or reinforce against an enemy success. In ground war, the general American approach has been the latter, and the Soviet approach the former. The Soviet approach is particularly well suited for fast offensives, while the

American approach is more defensive (even as part of an offensive).

In the heat of battle, it is easy to lose perspective, to judge something far more important than it is, or devote more resources to its attainment than it is worth. Let us look at an example. At the end of 1942, the German 6th Army was encircled at Stalingrad. Hitler, mesmerized by the concept of holding territory, forbade it to break out to the rear. For a variety of reasons, the *Luftwaffe* undertook the job of supplying an entire encircled army by air. It was manifestly incapable of doing so—a point made at the time by senior *Luftwaffe* officers and ignored by Goering—but tried nevertheless. In the next two months, before Stalingrad fell, the *Luftwaffe* lost most of its transport fleet, a special unit of bombers crucial to the submarine campaign, its bomber and instrument schools (aircraft and instructors were used at Stalingrad), and its prestige.[6]

These losses, because they could not be made good, were of far more consequence than the German 6th Army. The German high command had failed to think the problem through and paid a terrible price for nothing. The moral is clear: Make a cold, rational calculation of risks and rewards before committing to any operation.

Early in this chapter, we discussed the need to know oneself. That precept also is applicable in the area of training. Throughout this book, we have talked about mass and concentration. Mass and concentration require large formations in the air, formations that are not easy to plan, direct, or fly without extensive practice in peacetime. The idea is simple: If something is going to be done in war, it ought to be practiced in peace. If it has not been practiced, losses are likely to be high and the plan is unlikely to go as expected.

Command and control are necessary to bring the elements of air power together into a coherent fighting organization. The commander can use a system of explicit top down orders, or he can issue broad mission orders. Either system can work, as long as three key requirements are met: Officers and men from top to bottom must know what the system is and is not; it must have been practiced extensively in peacetime; and lower echelons must be given at least the minimum information required to carry out their responsibilities.

With this brief thought on command and control, we conclude this chapter. It does not include everything a commander needs to know to produce a winning air campaign. But it does include general principles that will get him started in the right direction.

The rest is up to him.

The Air Campaign
in Retrospect

Our purpose in this book has been to think through the problems confronting an air commander or staff officer in preparation for planning or executing an air campaign. A successful campaign clearly was contingent on a good plan, and construction of a good plan required a good understanding of the forthcoming action. The place to start was at the beginning.

Central to our thesis is the idea that air superiority is crucial, that a campaign will be lost if the enemy has it, that in many circumstances it alone can win a war, and that its possession is needed before other actions on the ground or in the air can be undertaken. Given that thesis, outlining the various situations that might obtain at the start of a campaign is necessary.

The situation facing a commander could at worst be one in which his bases were under air attack from the enemy while he had no capability to respond in kind. In this case, he would have no choice but to fight defensively, the worst way to fight an air war.

At the other end of the scale, the enemy's bases became subject to attack while those of our commander were safe. In between were the situations in which both sides had vulnerable rear areas or both were unable to reach the rear, so were constrained to fight over the front. Finally, in an anomalous case was the situation where combat air was not being used by either side.

Classification for its own sake may be academically interesting, but it is not militarily useful unless it leads to better operations. In the case of the air war, it does. In examining various cases, we saw that the ground relationship between defense and offense is reversed for aircraft. That reversal means that the air commander forced on the defense has a much tougher time than one might imagine, if his frame of reference were the ground. It also suggests that a commander should rarely accept the defense, if he has an offensive option. We also saw that several possible centers of gravity could be attacked to win air superiority, but that not all were available in every case.

Our examinations led us to the conclusion that numbers are important. In fact, they are so important that a primary goal of the operational commander ought to be to make sure that his forces outnumber the enemy every time they meet. The concept of fighting superior and winning followed—but with the caution that numbers did not mean theater numbers. Rather, the numbers that concerned us were the numbers that came together for an actual engagement.

We noted that the larger force almost always inflicts greater absolute and relative casualties on the smaller force. And it also usually suffers less in the process. This concept is certainly not new; in fact, it has been around for centuries in surface warfare. It is useful to know, however, that it is even more applicable in the air. Of course, qualitatively superior aircraft that are committed in battle in such a way as to be also superior in number to the enemy will accomplish far more than equal or inferior aircraft could hope to do.

After identification of the type of war and appropriate steps to win air superiority, we moved on to look at air interdiction and close support. We noted that destroying enemy equipment at or close to the source was more efficient than destroying them directly on the front. Thus, interdiction seemed theoretically preferable to close support. Recognizing that close support nevertheless was a vital air mission, we suggested that this scarce resource should be committed where the ground commander would commit his last division or artillery brigade—his operational reserve.

Our next subject was one relatively new to air operations—reserves. With few exceptions, the concept of reserves has been foreign to air forces. The theory of reserves, their ability to create a new situation and to shock and confuse the enemy, seemed as apropos to air as to ground campaigns. Our tentative conclusion, based on limited historical experience, suggests that air reserves ought to be maintained and committed at decisive points in the campaign.

We don't tend to think of war in the same terms as we think of music and concertos. But our discussion carried us to the conclusion that war plans had to have defined objectives and identified key forces if they were to lead to victory. The score for the concerto of violence had to be in consonance with the nature of the enemy, one's own nature, and the nature of the war. Discordance leads to defeat.

Lastly, we tried to integrate everything to produce a coherent air campaign plan. We saw that committing everything in emergency situations could be dangerous, if the commitment did not lead directly to a decision. We also saw that ground and naval forces could contribute to winning the air superiority that is vital to all. Finally, our discussion ranged to the use of deception and psychological warfare. Through it all ran the thread of concentration and mass.

Of all mankind's activities, war is the most baffling and intriguing. It brings out the best in men; and it uncovers the worst. War is the last argument of kings; appeal from its verdict, frequently impossible, is always difficult. War demands from its leading participants the coldest calculation, the most rational thought. Leaders lacking the ability to think clearly and precisely under war's enormous pressures pay dearly—often with their lives, always with the present and future of their followers. Methods of war change, but the principles of war—the essence of war—have not changed since Miltiades repulsed the Persians on the Plains of Marathon.[1]

War affects every person and nation it touches. The only way to mitigate its effects is to understand it thoroughly. Our purpose in this book has been to help in that process.

Epilogue
The Gulf War in Concept

John A. Warden III

As mentioned in the new foreword, one of the major refinements to *The Air Campaign* in the three years after its publication was integration of the centers of gravity ideas earlier discussed into a comprehensive view of the enemy as a system. In the Gulf War, we used this concept and many ideas drawn directly from *The Air Campaign* to plan the war and its execution.

With the concept of Iraq as a system and with the advent of precision weapons and stealth aircraft, we could think not only about creating wide system effects, but could also think about attacking Iraq as a system in parallel instead of in the serial fashion which the old era suggested. The difference between serial and parallel, between system war and military war[1] are crucial but hard for many to grasp. In fact, the differences are so extreme that many of the terms which were reasonable with serial and military war are utterly dysfunctional when applied to parallel and system war.

First, the purpose of war ought to be to win the peace that follows and all planning and operations should be directly connected with the final objective. Although we pay lip service to this idea, in policy, military, and academic worlds, we easily get lost

---

[1] I am using military war to denote an approach to war where the most important thing becomes the Clausewitzian fixation on the clash of armies.

in a Clausewitzian world where defeat of the enemy military forces becomes an end in itself rather than merely one of a number of possible means to a higher end.

The proposals made to General Schwarzkopf on the 10th of August 1990 flowed from a very specific view of the peace that should follow a war with Iraq and from an understanding that attainment of our peace objectives depended on our recognition that Iraq was and is a complex system. The plan put forward was to attack Iraq so as to change Iraq the system such that it would be compatible with the envisioned post war peace.

In the simplest of terms, the post war peace had to contain these two major elements: Iraq out of Kuwait and an Iraq that would not be a strategically threatening regional superpower for an extended period of time(General Schwarzkopf agreed that about a decade would be adequate). We could achieve the latter by destroying Iraq as a state, but it was clear that the resultant power vacuum might be even more of a threat to regional stability than a powerful, belligerent Iraq.

The plan to produce this postwar peace began with an analysis of Iraq based on the Five Rings system (an extension of the centers of gravity ideas first laid out in *The Air Campaign*); we had developed and debated this concept in the Air Staff over the two years prior to the Iraqi invasion of Kuwait. The underlying assumption of this analytic approach is that all organizations are put together in about the same way. Thus, every organization has a leadership function to give it direction and help it respond to change in its external and internal environments; every living entity has an energy conversion function to take one form of energy and convert it into a different kind of energy; each has an infrastructure to hold it together; each has a population; and each has fielded forces to protect and project the organization. To make it easy for everyone to visualize the concept, we laid it out in the graphical form of five concentric rings with the leadership ring at the center.

By putting the five rings together graphically, we immediately grasp the idea that we are dealing with a system, that the military—or fielded forces—are but one part of the system and the most peripheral—and that the leadership ring is of central importance.

In the case of Iraq, our goal was to reduce the energy level of the entire system enough to reach our peace objectives. With this approach to strategic analysis, we always begin our thinking in the center; only at the center can a single input of energy (an entreaty from the President of the United States, or something physical like a bomb) result in a significant change in the system. On occasion, the single input of energy has led to the collapse of empires—for example, Darius III's departure from the fields of Arbela led almost instantly to the transfer of Persia to Alexander. For the most part, however, this doesn't happen and it would be a poor strategist who bet everything on it. Remembering that our goal was to affect the entire Iraqi system, we identified additional centers of gravity progressing from the inside to the outside. The version we presented to General Schwarzkopf at our second meeting with him a week after the first looked roughly like this in simplified form.

Iraqi Target Systems

Leadership	Key Production	Infrastructure	Population	Fielded Forces
Saddam Hussein Government	Electricity	Railroad Bridges	Military Elites	Strategic Air Defenses
National Communications	Retail Petroleum		Foreign workers	Strategic Offensive Systems (air and missile)
Internal Security Forces	Weapons of Mass Destruction		Baathists	
			Middle Class	

It is important to understand that the five rings and the table of Iraqi centers of gravity are describing a system. With the understanding of Iraq as a system our task becomes one of converting it into something that will be in consonance with our post war objectives. The faster we can force the conversion, the more likely we are to succeed, for the slower we proceed and the more serially we approach the problem, the more likely it is that the enemy will find ways to counter our operations. Thus, our goal was to bring the Iraqi system under rapid—or parallel—attack. And for the first time in the history of non-nuclear warfare, we had the concepts, aircraft, and weapons to make parallel attack possible. A quick comparison with World War II will be illustrative.

In World War II, the United States began the daylight bombing of Germany in January of 1943. Its principal bomber, the B-17, had over the course of the war about a one thousand yard Circular Error Probable—or CEP.[2] To put this in perspective, if you want to hit something about a third the size of football field and you want a 90% probability of at least one bomb falling into the described area, you must drop over nine thousand bombs. In WWII terms, that meant flying a thousand B-17 sorties and putting about ten thousand men at risk over the target. Because of the relative inaccuracy of weapons, it was necessary to attack large complexes instead of the important parts of them. Likewise, it was necessary to group large numbers of airplanes together for two reasons: first, they had to have sufficient mass to penetrate enemy defenses, and second, it was necessary to drop a very large number of bombs in order to have any chance of hitting anything.

When the United States began its daylight operations against Germany, it could only put relatively small numbers of bombers in the air at any time, and for the reasons noted, could only attack one target per raid. The result was a serial attack to which the Germans responded by repairing damage and improving their defense schemes. For the latter, the Germans

---

[2] CEP: the radius beyond which half the bombs dropped will fall.

put an enormous amount of resources into building and manning antiaircraft guns and they also withdrew dangerously large numbers of fighters form the tactical fronts. The strategic base of Germany was so important that Hitler and his high command recognized that they had to try to defend it regardless of the cost. This historical fact, which has repeated itself in every instance where a state has found itself under strategic attack of any kind, rather clearly shows that government and military leaders understand the importance of a secure strategic base far more than a disconcertingly large number of academic commentators.

In the German case, Albert Spear recognized in 1943 that strategic bombing would doom his country. Spear was able through heroic efforts to push back the final collapse of the transportation and energy systems until early 1945; by 1945, there were so many bombers attacking Germany's strategic centers that the damage accumulated faster than the Germans could deal with it.

If our tools in the Iraq case had been similar to those available in World War II, we would have been compelled to attack Iraq serially and we would have started with some small part of its air defense system. If we were very lucky, after a long period of time, we might have been able to start the reduction of the key inner rings but that would have been far into the future. Likewise, it would have been chancy; Iraq in the summer of 1990 had perhaps the most modern air defense system in the world.

Fortunately, we had completely different tools to use against the Iraq system. We had an information system that allowed us to coordinate operations ranging from Colorado Springs to Great Britain to Baghdad itself. We had stealth aircraft that penetrated by themselves and thus made it possible to bring many targets under simultaneous attack. We had unmanned missiles. And most important of all, we had bombs that had a very high probability of hitting that against which they were aimed. Precision has changed the face of warfare.

To update our World War II example: in the Iraq war, if we wanted a high (90%) probability of hitting a target a third the size of a football field, we could confidently dispatch one F-117 stealth fighter manned by one pilot who would drop one of his two bombs. This represents an incredible four order of magnitude improvement in accuracy and personal productivity over World War II. It also does something else of great interest. In World War II, conventional bombing seemed to do a lot of damage and indeed it leveled entire cities. Despite the apparent damage, however, vital functions continued in many cases. In 1945 Berlin, for example, the telephone and teletype systems continued to work until the very end as did the water system and even much of the electrical system. This was in a city that from the air looked badly hurt. The reason for this anomaly is simple: the important things tend to be small and the odds are good that they won't be hit directly. In other words, even though the strategic target base is relatively small (several hundred targets for even large nations), it is very difficult and time consuming to affect with inaccurate weapons and serial operations. Contrast 1945 Berlin with 1991 Baghdad: within minutes of the start of the war, electricity is off in Baghdad and does not return until after the war's end, and the ability to communicate plummets.

In the old inaccurate world, we evaluated the effects of bombing or artillery barraging in terms of the physical damage done to the target. In the Gulf War, intelligence analysts—and subsequent commentators—would note that only ten percent of the electrical facilities were destroyed or only fifteen percent of the road surface between Baghdad and Basra, or only twenty-five percent of the communications facilities. They would then extrapolate their damage observations to equate with function and say then that the Iraqis still had 75% of their capacity available which was more than enough. The world, however, is different in the era of precision and parallelism. With the two, the targets become the important parts of the electric, communications, or transportation system. In the case of the latter, destruction of some thirty bridges (a small portion of the road surface) between Baghdad and Basra reduced movement by nearly a hundred percent and dropped the flow of critical supplies into Kuwait below the survival level in the first three weeks of the war. The obvious result was that the Iraqis in Kuwait were quickly depleting what they had stockpiled. The electrical and communication systems were similar: by hitting the key facilities in each, the output levels dropped precipitately—even though physical damage was relatively minor. In the past, war efforts tended to be aimed at physical structures and success measured accordingly. Today, war efforts aim at function and we are successful when function stops regardless of physical damage. Failure to understand the shift from the physical to the function has significantly obfuscated analysis of the Gulf War and has led many writers to erroneous conclusions.

In consonance with *The Air Campaign* principles, the plan was based on offensive operations; the overall thrust of operations in the first hours of the war was to begin inducing a strategic paralysis in Iraq which would simultaneously begin the process of cutting Iraq down to an acceptable postwar size and making it impossible for Iraq to do anything about it. The operations were parallel, but the limitations of the printed page force us to explain what we did and why in a serial way. For convenience, we will start at the center of the five rings and work to the outside. *The Air Campaign* reader will note that most of the centers of gravity mentioned earlier in the book came under attack. The difference was that we now understood their relationship to the entire nation and so could better choose which ones and in which order to strike.

At the very center was the Saddam Hussein government that, of course, included Saddam himself. In our first meeting with General Schwarzkopf we stressed the importance of making clear to the Iraqis and to the rest of the world that our problem existed because of Saddam Hussein's policies—not because we hated Iraq or Iraqis. By presenting the war in this way, we wanted also to make clear to the Iraqis that they would fare well in a postwar world that didn't include Saddam. One of the marvels of parallel war is the ability to conduct simultaneous attacks on multiple fronts and targets without causing the disaster that such dispersion of effort would have produced in the past. Indeed, not to make this case would have been naive and would have caused us to lose our focus. Likewise, we discussed with General Schwarzkopf what would happen in the event that Saddam was not displaced. Our view, with which the General agreed, was that it would be far better for Iraq if Saddam was no longer in power. However, as long as we had taken away from Saddam the tools (the system) he needed to be a regional superpower threat, his disposition was not of overwhelming importance.

The consensus of the intelligence community in August 1990 was that Iraq was on its current course of conquest primarily because of the ambitions of Saddam Hussein and that if he were gone, Iraq's policies at least in the short term would be less bellicose. Further, the majority of analysts with whom we consulted (inside and outside of government) were of the opinion that no one who might replace Saddam would be as bad as he simply because they would lack the power base it had taken him almost two brutal decades to build. Thus, if we could create a situation that led directly or indirectly to Saddam's departure from office, we would have contributed to our postwar peace objectives of creating an Iraq less threatening to its neighbors. In addition to the geostrategic rationale, there was a second reason that we needed to attack the Saddam government: to decrease its ability to oppose our operations against Iraq. Strategic direction and coordination of combat, psychological, and support operations above a tactical or medium operational level come from national capitals regardless of the type of government. Furthermore, when damage is accreting rapidly—as it does it parallel war—decisions must come much faster than in peacetime.

We could picture Saddam's government organization the same way we could picture Iraq as a whole; the Five Rings approach recognizes fractal relationships that repeat themselves from the very large to the very small. In other words, each part of the system is defined by its own Five Rings structure right down to the level of an individual. With this picture of the government, we could think about attacking it in parallel. We clearly didn't know where all the major offices were, but we were able to discover the locations of many and to attack each with a single bomb. We expected to hit most command centers of the government related to running the country in war. We knew that each one of these facilities had a backup. We also knew that the backups in general would not be quite as well equipped and manned as the primaries. And here the reader should begin to get a sense of the effect of massively parallel attack.

In our everyday experience, we know how much our efficiency falls off when we lose our telephones for a day, or when we change our office location or telephone number. Despite our best efforts, people don't find us, we can't put our fingers on the important paper we had yesterday, and so on. In normal affairs we accept these problems knowing they will eventually go away. Imagine what happens, however, when the majority of a government's senior leaders and their key staffs change their office addresses and phone numbers without notice in the midst of a very stressful situation. Does anyone really believe that the efficiency of the government must not decrease significantly and rapidly? In the days of inaccurate weapons, forcing a major change on the physical manifestations of government was simply not possible; in the Gulf War it was possible and it happened. Attacks on government command facilities had another effect; senior officials simply had to become more mobile in order to reduce their chances of being killed. Saddam Hussein himself was reluctant to use even a cellular phone for fear that he would be targeted. Picture the President of the United States and most of his key advisors incommunicado for extended periods each day and ask how well and quickly the government will make critical decisions. And all this happens despite an apparently low level of physical destruction.

Attacks on the government communications system accelerated the breakdown in government efficiency, reduced significantly the very high volume of communications the Iraqi military high command customarily used to direct field operations, and also made it more difficult for Saddam Hussein to communicate directly with his citizenry.

Many observers thought the Iraqi military didn't need much in the way of command from Baghdad, but in fact they had developed a very effective system to convey high volumes of information to the front. Commentators who have never participated in military exercises or operations frequently confuse tactical, operational, and strategic levels of command. They think that because a small unit (a company for example) is given a degree of autonomy, that the same thing must apply to larger units. It doesn't—anywhere—simply because offense or defense at an operational (theater, for simplicity) level requires enormous coordination to ensure that logistics, supporting fires, communications, deception, logistics, and the rest come together at the right time and place. Few countries are capable of making this come to pass. So far, those that have done it on an operational level have only been able to do it with very high bandwidth communications. The Iraqis had developed their system during the Iran war and had become quite proficient with it. They, of course, had spent great sums of money to buy the most modern equipment and had bought enough to have a very robust and redundant system. It would have served them well against any enemy except one equipped with precision weapons and the parallel war concept.

As mentioned, another facet of the attack on strategic communications was the desire to make it more difficult for Saddam to talk to fellow Iraqis. As we all know, a harsh dictatorship is highly dependent on keeping its image and presence in front of the people. If it seems to disappear, people begin to behave as though it were not there—which was one of our many parallel goals. There is more to this equation than just taking away communication from the dictator; the attacker must provide a substitute. At our first briefing to General Schwarzkopf, we said very explicitly that the strategic psychological operations campaign was entirely as important as the bombing campaign. Unfortunately, despite the best efforts of a lot of people including General Schwarzkopf, there was never a real strategic psychological operations campaign which would have done even more to facilitate possible coups or other Iraqi actions against Saddam and his Tikrit clan.[3]

Saddam's security forces were the next area of attention. These comprised the various KGB and Gestapo like services that protected Saddam and which carried out his reign of terror. We believed that attacks on these groups would loosen Saddam's grip and again facilitate coups or other direct action against Iraq. Keep in mind that this was but one of many parallel attacks and the success of the overall campaign did not depend on its success. As will be clear shortly, if not already, our air strategy was not a "decapitation" strategy, although attacks against the leadership were very much a part of the plan.

At the time of the war, we called the next ring out "Key Production;" the idea was to highlight those activities on which the country as a system totally depended. Although we subsequently changed the name to "System Essentials," the concept is the same. In this ring, we identified electricity, retail petroleum, and weapons of mass destruction research and manufacture as the energy conversion functions that were vital to Iraq being what Saddam wanted it to be.

People who fail to understand the systems and center of gravity concepts failed to grasp the importance of electricity as a strategic target. It is, in virtually every country that has it, a strategic target of the first importance. Electricity is the most efficient way to move energy around a country. It powers everything—from radar antennas to elevators to telephone

---

[3] There was an effective tactical psychological operations campaign against the Iraqi army in Kuwait. Unlike the strategic campaign which needed Washington direction and participation, the tactical campaign was under General Schwarzkopf's authority and he did a great job with it

switching centers to computers. It is true that some of these have back up generators—but the backup generators are precisely what their name implies. They are not designed to be long term sources of electricity and they do not substitute well for primary electricity even for short periods. The logic is pretty simple; if backup generators were as good as primary electricity, there would be no national grid. When the electricity in a country goes off, it immediately puts a strain on almost every activity in a country and forces the occupants to expend energy to find alternatives. Thus, by shutting down the electrical system, with relatively little effort, we were able to affect almost everything and everyone in most of Iraq.

Consider a trivial example. Most of the government buildings in Iraq were multistory which meant they had elevators. When the power went out, the elevators in many cases stopped functioning at their previous level so occupants of upper stories had to walk up the steps. No big deal, one might say, until you realize that this simple act has probably imposed a five minute time penalty on every government worker and has also made him look for excuses not to go to his office or leave it.[4] Although it has much more important effects, shutting off electricity is rather like pouring a layer of molasses over the whole country; people can still move, but they move slower and they spend energy they would otherwise have put to more profitable uses. Attacks on electricity were exceptionally valuable in creating the system wide strategic paralysis that we wanted to impose on Iraq.

Our attacks on retail petroleum had a similar rationale: create a major problem that would have an effect across the country. Here again, many commentators fail to understand the impact of stopping the production of refined petroleum products. They become mesmerized with the idea that it is only the fielded forces which count and the problem for the affected country is nothing more that a reallocation of resources. In fact, it is far more because of all the demands for the resources. The

---

[4] In the late 1970's, the Carter administration shut down the escalators in the Pentagon in order to conserve energy. The amount of important physical interaction among occupants of the six levels of the Pentagon fell noticeably!

user level shortages, which develop quickly because it takes time to deal with a problem that arises overnight, magnify the problems caused by the failure of the electric system. Every backup generator has a relatively small supply of fuel collocated with it which needs replenishment when the backup generator is used for more than a few hours or a day. With shortages in the fuel supply, however, finding generator fuel and getting it into the tank becomes difficult or impossible. Further complicating the reallocation problem is the fact that much of the support for even the military comes through private or non-military transportation or supply companies. Given a lot of time to respond to a gradually developing problem, a competent bureaucracy can develop an effective response. However, it is not clear that there is any bureaucracy in the world that can solve such a challenging problem when telephone coordination with multiple agencies is impossible and physically finding the people who can make decisions is likewise difficult or impossible.

In the interest of brevity, we will move to the fielded forces ring. Note that the centers of gravity in this area we had identified for the strategic part of the campaign were just air defense and offense. We saw them as being a part of the overall system that we had to convert in order to achieve our objectives. Reduction of the air defense system allowed us to use all of our attack aircraft without fear of large losses. In addition, its loss put Saddam in a precarious position for his future and that of his country fell into the hands of his attackers as soon as he was unable to defend himself and Iraq. Operations against Iraq's strategic air offense (air platforms and missiles) were necessary to deprive Iraq of its ability to conduct potentially dangerous strategic counteroffensives. Most worrisome in early August, and a subject of conversation with General Schwarzkopf, was the likely outcome of serious Iraqi attacks on Israel. We were confident that we could suppress the air threat, but there was no obvious direct way to prevent the Iraqis from launching mobile missiles. Our goal was to do as much as we could indirectly to lessen the numbers and types fired and to ameliorate their effect as best we could.

During the Iran war, the Iraqi Air Force had conducted very sophisticated, very long-range operations against such key Iranian targets as oil tankers and petroleum facilities. Before the Gulf War, everyone had a healthy respect for the Iraqi Air Force and was most concerned. From a planning standpoint, it was necessary to devise a way to neutralize this very real threat. Like the rest of the operations, our plan was parallel: knock out the air defense command and control system and reduce national communications quickly in order to drive individual air force units into autonomous operations. The result would mean they would have to deal with a national threat with only local information available to them. The consequences were severe and quick to become obvious; Iraqi pilots took off without any picture of the air defense system. They didn't know whether an enemy fighter was orbiting high over their field or not. Too frequently there was and their efforts to fly were futile. Their reaction was entirely correct; they took temporary refuge in the best air defense shelters in the world—shelters they reasonably thought would be proof against anything other than a direct nuclear hit. Much to their amazement (and to the amazement of many American officers), the latest generation of precision bombs were highly effective against these shelters. Thus, the Iraqis died if they flew and died if they didn't. Note that the proposals in *The Air Campaign* about the best ways to win air superiority and put the enemy air force out of action turned out to be quite accurate and useful.

In our first discussion with General Schwarzkopf, we suggested that it was possible to secure our objectives without attacking the Iraqi army in Kuwait. The next day, we met with General Colin Powell, Chairman of the Joint Chiefs of Staff; he said he wanted to attack the Iraqi army to send a political message to Saddam and others. Of some interest, many in August 1990 (and even later) wanted only to attack the Iraqi army in

Kuwait; after all, was not that the problem and should not one solve problems where they manifest themselves?

The direct solution called for application of standard Army AirLand Battle doctrine which would have meant air and artillery attacks to soften the Iraqi army in Kuwait followed by American and Coalition ground attacks to push the Iraqis out of Kuwait. Doubtless, it was within the theoretical military capability of the coalition to take this approach given a year or more to build up the attacker to defender ratios military commanders would have demanded. On the other hand, the President of the United States would have had great difficulty in securing political support for an operation which would have led to very high casualties; forecasts made in the late summer of 1990 under the aegis of the Joint Chiefs of Staff estimated 20,000 American casualties.

Put aside for a moment the questions about logistical and political feasibility and assume the hypothetical operation was successful and that Saddam had lost as much of his army in Kuwait as he lost in the actual operation. At the end of a roll back war that involved no serious attack on the interior, Saddam would have been out of Kuwait, but would have suffered no significant strategic damage. The war would likely have ended with a cease-fire with two substantial forces facing each other across the Kuwait-Iraq border. Saddam's losses would have been much less than they were in the war with Iran, and despite losing up to half of what he had in Kuwait, he still would have had one of the most capable armies in the world. Under these circumstances, does anyone believe that he would have permitted the gross infringements on Iraqi sovereignty that followed the war and are still occurring? Would he have permitted unarmed UN inspectors to wander around Iraq commanding the destruction of important and expensive weapons and programs? To say the least, it seems unlikely. Saddam's ground force losses were trivial compared to what he had previously weathered and what he was willing to accept as part of his own strategy of attrition. In the real world, it was the terrible weakening of Saddam's real center of gravity, his strategic base, that forced him to accept unconscionable (from his perspective) peace terms and that has allowed a handful of American airmen to keep him within the bounds the United States defined (which may well have been far less than what we could have demanded).

As noted above, the Gulf War confirmed many of *The Air Campaign* concepts. Several more follow in condensed form:

> Defense against air attack is extraordinarily difficult and has been since the earliest days of aviation. It is doubtful that any country in the world could have countered the air offensive that hit Iraq in the early morning hours of 16 January 1991.

> The Iraqis lost air superiority in the first minutes of the war—and were doomed thereafter. No one can afford to lose air superiority!

> General Schwarzkopf identified and used airpower (from all services) as the "key force" for the first 38 days of the 41 day war.

> It was entirely feasible to bypass "the castles" and go directly to the capital. (Recall Chapter 9 "The Orchestration of War.")

> Air superiority is key to strategic operations, just as it is key to every other kind of operation. Winning air superiority is difficult and one of the surest ways to fail is to think you can take the parsimonious approach and just go for local superiority. Local air superiority is a very dangerous idea simply because it ends up requiring air defense which is very difficult.

> It wasn't necessary to conquer territory in order to defeat the Iraqis.

Mass and concentration are as important as ever, but stealth and precision have allowed us to understand these concepts in terms of their effects, not on the number of people or machines committed.

The effects of attacks on command and control centers of gravity were even better than anticipated in *The Air Campaign*.

There were many people who made *The Air Campaign* against Iraq such an enormous success. President George Bush and General Norman Schwarzkopf were obviously the most important; both provided the extraordinary strength, commitment, and leadership needed to rally the world and execute a radical plan. Also well known are officers like General Chuck Horner who ably orchestrated the air forces under his command and held to the right course of action despite great pressure from more conventional air, sea, and land officers who wanted to revert to the old way of fighting. Less well known are a handful of young officers who helped develop the concept of *The Air Campaign* on the Air Staff in the late 1980's and who then played key parts in planning and carrying out the Gulf War. Among the standouts were Dave Deptula who assumed the key substantive planning role in Riyadh, Ben Harvey who was integral to the targeting effort, and Ronnie Stanfill who helped educate the American intelligence community on the new forms of war.

Since the war ended, there has been much discussion about whether a revolution in military affairs took place. In my view, the answer is unequivocal; the Gulf War was the first conflict in the first true military technological revolution in history. As with all revolutions, whether in war or in business, the majority of people have difficulty accepting the reality of revolution and thus miss the huge opportunities available to them. Such is certainly the case with respect to the Gulf War; when the lessons of the war are finally understood and accepted, those who have learned the lessons will benefit enormously while those who continue to deny the fact of change will suffer the consequences.

Endnotes

AIR CAMPAIGN IN PROSPECT

1. One exception is Richard E. Simpkin's *Race to the Swift. Thoughts on Twenty-First Century Warfare* (London. Brassey's Defence Publishers, 1985), a very useful work on the theory of operations in the ground campaign.

2. The Battle of Cannae (216 BC) was a major clash near the ancient village of Cannae, in Apulia, Italy, between the forces of Rome and Carthage during the Second Punic War, in which the Carthaginian general Hannibal won a great victory. The eventful campaign was begun by a new, aggressive move by Rome. An exceptionally strong field army, estimated at between 48,000 and 85,000 men, was sent to crush the Carthaginians in open battle. On a level plain near Cannae, chosen by Hannibal for his battleground, the Roman legions attacked. Hannibal deliberately allowed his center to be driven in by the Romans' superior numbers, while Hasdrubal's cavalry wheeled round to take the enemy's flank and rear. The Romans, surrounded on all sides and so cramped that their superior number aggravated their plight, were practically annihilated.

3. Carl von Clausewitz, *On War*, transl. and ed. by Michael Howard and Peter Paret (Princeton, N.J.. Princeton University Press, 1976), p. 595.

4. J.F.C. Fuller, *The Generalship of Alexander the Great* (London: Eyre & Spottiswoode, 1958), pp. 95–103.

5. Fuller, *The Decisive Battles of the Western World, Vol. 2* (London: Eyre & Spottiswoode, 1965), pp. 37–39.

6. Direct attack on the enemy's power base—bypassing his fielded military forces—is also an option. See chapter 9.

CHAPTER 1

1. See subsequent discussion in this chapter on Vietnam—the one example that on the surface could be seen to contradict these general principles.

2. Cajus Bekker, *The Luftwaffe War Diaries*, transl. and ed. by Frank Ziegler (New York: Ballantine Books, 1969), p. 31.

3. Williamson Murray, *Strategy for Defeat. The Luftwaffe 1933–45* (Maxwell Air Force Base (AFB), Ala.: Air University Press, 1983), pp. 36–37.

4. Ibid., p. 86.

5. Telford Taylor, *The Breaking Wave* (New York. Simon and Schuster, 1967), p. 71.

6. Ronald Lewin, *Rommel: As Military Commander* (New York: Ballantine Books, 1972), p. 275.

7. Ibid.

8. *The Impact of Allied Air Interdiction on German Strategy for Normandy* (Washington, DC. US Air Force Assistant Chief of Staff, Studies and Analysis, 1969), p. 14.

9. Generalleutnant Klaus Uebe, *Russian Reactions to German Airpower in World War II* (Maxwell AFB, Ala.: Aerospace Studies Institute, 1964), p. 100.

10. William W. Momyer, *Air Power in Three Wars (WWII, Korea, Vietnam)* (Washington, DC: US Air Force, 1978), p. 117.

11. Randolf S. and Winston S. Churchill, *The Six Day War* (Boston: Houghton Mifflin Company, 1967), pp. 86, 177.

12. The Insight Team of the London *Sunday Times*, *The Yom Kippur War* (Garden City, N.Y.: Doubleday & Company, Inc., 1974), pp. 161, 204. The London *Sunday Times* Insight Team of reporters combines first-rate investigative journalism with lively writing and opinion. The team has authored several books, including bestsellers *Philby, An American Melodrama*, and *Do You Sincerely Want to be Rich?*, and *Watergate*. *The Yom Kippur War* derives from the extensive coverage of the fourth Arab-Israeli war in the London *Sunday Times* during October 1973.

13. A.J.C. Lavalle, ed., *The Vietnamese Air Force 1951–75: An Analysis of Its Role in Combat* (Washington, DC. Superintendent of Documents, US Air Force Southeast Asia Monograph Series, Vol. 3, Monographs 4–5, 1976), pp. 58–59.

14. *The Yom Kippur War*, pp. 161, 204.

15. Taylor, pp. 108–10.

16. Bekker, pp. 348–9.

17. *The Yom Kippur War*, p. 213.

18. During this period, the few British raids on the German homeland had no military effect on the battle—although their subsequent political effect perhaps was significant. More on this aspect when deception is discussed.

19. Richard Suchenwirth, *Historical Turning Points in the German Air Force War Effort* (Maxwell AFB, Ala.. Air University, 1959), pp. 66–67.

20. Taylor, pp. 150–51.

21. Bekker, pp. 313–18.

22. Murray, p. 285.

CHAPTER 2

1. Our emphasis is on the use of air to defend against air because experience to date has shown that ground-based defenses, whether antiaircraft artillery or guided missile systems, have not been able to provide effective opposition to an air offense. In 1973, Syrian missile defenses on the Golan Heights forced the Israelis to stop their air attacks for several hours, after which time the Israelis destroyed the missiles and the positions the missiles were to protect. This 1973 ground-based defense success is the best so far recorded. This observation is not meant to say that ground-based defenses can be ignored by the attacker, or that they have no utility for the defender. For the latter, they may help channelize enemy air operations or may at least force the enemy to devote some effort to overcoming defenses. In the absence of defenses, he might apply that effort in a more dangerous fashion. At some point, perhaps with the introduction of directed-energy weapons, ground-based defense may succeed in thwarting the offense for an extended period. The operational commander must be alert to that possibility: Technological advantages can be crucial.

2. Adolf Galland, *The First and the Last* (New York. Ballantine Books, 1963), p. 137.

3. Clausewitz, pp. 357–59.

4. Suchenwirth, pp. 117–18.

5. D. Clayton James, *The Years of MacArthur 1941–45* (Boston: Houghton Mifflin Company, 1975), pp. 207, 227.

6. James, p. 281.

7. Ibid., pp. 197–99.

8. George C. Kenney, *General Kenney Reports* (New York: Duell, Sloan, and Pearce, 1949), p. 324.

9. James, p. 324.

10. Kenney, pp. 276–78.

11. US Strategic Bombing Survey, *Japanese Air Power* (Washington, DC: Military Analysis Division, 1946), p 14.

13. Kenney, p. 241.

14. *Japanese Air Power*, p. 15.

15. US Strategic Bombing Survey, *Air Campaigns of the Pacific War*, p. 60.

16. *The Yom Kippur War*, pp. 161, 167.

17. Ibid., pp. 213, 238.

18. Ibid., p. 204.

19. Bekker, p. 229.

CHAPTER 3

1. The V-1 and V-2 missiles had little military impact. Had there been enough of them, had they been more accurate, and had the Germans concentrated them against the ports and airfields, they might have been more significant.

2. To illustrate, less than 1 percent of American military pilots have become aces (shooting down five or more enemy aircraft), but that 1 percent has accounted for more than 30 percent of all enemy aircraft destroyed in the air. Gene Gurney, *Five Down and Glory* (New York: Ballantine Books, 1957, 1965), pp. 207, 242.

3. Suchenwirth, p. 83.

4. Bekker, p. 312.

5. Ezer Weizman, *On Eagle's Wings* (New York: MacMillan Publishing Co., Inc., 1976), p. 223, and Churchill, p. 86.

6. *Japanese Air Power*, p. 14.

7. Haywood S. Hansell, Jr., *Strategic Air War Against Japan* (Maxwell AFB, Ala.: Airpower Research Institute, 1980), pp. 76–80, and Albert Speer, *Inside the Third Reich*, transl. by Richard and Clara Winston (New York: Avon Books, 1971), pp. 365–67.

8. During American planning and execution of the bombing campaign against Germany, some of the planners maintained that destroying enough single-target systems would win the war. Critics of this approach disparagingly referred to these target systems as "panaceas." In retrospect, the petroleum, transportation, and electrical generating systems might have come close to qualifying as real "panaceas." See the remainder of this chapter for more detail.

9. A.J.C. Lavalle, ed., *The Tale of Two Bridges and the Battle for the Skies over North Vietnam* (Washington, DC. US Government Printing Office, USAF Southeast Asia Monograph Series, 1976), p. 151.

10. Murray, pp. 274–76.

11. Bekker, p. 7.

12. Ibid., p. 31.

13. Taylor, pp. 138–39, 151–59.

14. Murray, pp. 274–76.

15. Bekker, pp. 200–201, and Taylor, p. 145.

16. Bekker, p. 240.

17. Lieutenant Colonel James H. (Jimmy) Doolittle led 16 B-25 *Mitchell* medium bombers on a daring raid on military targets at Tokyo, Yokohama, Yokosuka, Nagoya, and Kobe on 10 April 1942 from the deck of the US Navy Aircraft Carrier USS *Hornet*. This first air attack on Tokyo in World War II, which carried the Battle of the Pacific to the heart of the Japanese Empire, bolstered US morale, slowed Japanese offensives, and earned Colonel Doolittle the Medal of Honor.

18. Gordon W. Prange, *Miracle at Midway* (New York: McGraw-Hill Book Company [Penguin Books], 1983), pp. 25–27.

19. *Japanese Air Power,* p. 10.

20. Murray, p. 282.

21. Benjamin S. Lambeth, *Moscow's Lessons from the 1982 Lebanon Air War* (Santa Monica, Calif.: Rand Corporation, 1984), pp. 4–7.

22. Paul S. Cutter, "EW Won the Bekaa Valley Air Battle," *Military Electronics/Countermeasures,* January 1983, p. 106.

23. Lambeth, p. 8.

24. A victory for the English, on 26 August 1346, in the first decade of the Hundred Years' War against the French. Edward III of England landed 4,000 men-at-arms and 10,000 archers (longbowmen) on the Cotentin peninsula in mid-July 1346, and ravaged lower Normandy west of the Seine and as far south as Poissy, just outside Paris. Philip VI of France advanced against Edward with 12,000 men-at-arms. Edward turned sharply northeastward, crossing the Seine at Poissy and the Somme downstream from Abbeville, taking a defensive position at Crecy- en-Ponthieu, where he posted dismounted men-at-arms in the center, with cavalry to their right and left, and archers on both wings. Italian crossbowmen in Philip's service began the assault on the English position, but were routed by the archers and fell back into the path of the French cavalry's first charge. More and more French cavalry came up to make further thoughtless charges at the English center. But while the center stood firm, the archers wheeled forward and the successive detachments of horsemen were mowed down by arrowshots from both sides.

25. Douhet (1869–1930), an Italian military officer, generally is regarded as the father of strategic air power. Trained as an artillery officer, from 1912 to 1915 he commanded the Aeronautical Battalion, Italy's first aviation unit. Through his efforts, the three-engine *Caproni* bomber was ready for use by the time Italy entered World War I. Douhet's theory of the important role of strategic bombing in disorganizing and annihilating an enemy's war effort was incorporated into future military plans of Italy and the United States. He further advocated creation of an independent air force.

26. Murray, pp. 6–14.

27. Hansell, pp. 18–19.

28. US Strategic Bombing Survey, *Summary Report (Pacific War)* (Washington, DC: US Government Printing Office, 1946), pp. 25–26.

29. Ibid.

30. Cutter, p. 106.

CHAPTER 4

1. The operational commander's duty is to ensure that he masses superior forces at a particular time and place. That he is inferior in the theater does not relieve him of this duty. In fact, it is the essence of generalship.

2. Wesley F. Craven and James L. Cates, *The Army Air Forces in World War II, Vol. II* (Chicago: The University of Chicago Press, 1949), pp. 704–6.

3 The Insight Team of the London *Sunday Times, The Yom Kippur War,* p. 161. In addition, note that the Israelis lost all of these aircraft to ground-based defenses. This action is the only known instance before or since in which such defenses had an effect on this scale.

4. *The Relationship Between Sortie Ratios and Loss Rates for Air-to-Air Battle Engagements During World War II and Korea—Saber Measures (Charlie)* (Washington, DC. Headquarters, US Air Force, Assistant Chief of Staff, Studies and Analysis, 1970), p. 15.

5. J.F.C. Fuller, *The Decisive Battles of the Western World, Vol. III* (London. Eyre & Spottiswoode, 1963), p. 471

6. Lambeth, p. 8.

7. John H. Morrow, *German Air Power in World War I* (Lincoln: University of Nebraska Press, 1982), p. 109.

8. Galland, pp. 150–1, 187.

9. "Ultra" was the code name for the Allied exploitation of material deciphered from the German Enigma coding machines.

10. Bekker, pp. 232–42, and Taylor, pp. 151–61.

11. Galland, pp. 30–31.

CHAPTER 5

1. Bekker, p. 242.

2. Ibid., p. 525.

3. Momyer, pp. 147–48.

CHAPTER 6

1. Clausewitz, p. 95.

2. Robert F. Futrell, *The United States Air Force in Korea 1950–53* (Washington, DC: Office of Air Force History, 1983), pp. 261–63.

3. William Manchester, *American Caesar: Douglas MacArthur 1880–1964* (Boston: Little, Brown and Company, 1978), p. 611.

4. Momyer, p. 168.

5. Suchenwirth, pp. 90–91.

6. David Irving, *The Trail of the Fox* (New York: Avon Books, 1978), p. 175.

7. Ibid., p. 92.

8. *USAF Tactical Operations. World War II and Korean War* (Washington, DC: USAF Historical Division, Liaison Office, 1962), p. 30.

9. F.M. Salagar, *Operation "Strangle" (Italy, Spring 1944): A Case Study of Tactical Air Interdiction* (Santa Monica, Calif.: The Rand Corporation, 1972), p. 62.

10. Ibid., p. 66.

11. Lewin, p. 274.

12. Anthony Cave Brown, *Bodyguard of Lies* (New York: Bantam Books, Inc., 1976), p. 429.

13. *Summaries of Selected Military Campaigns* (West Point, N.Y.: Department of Military Art and Engineering, US Military Academy, nd, Special Printing for the Department of History, US Air Force Academy, 1960), p. 142.

14. Murray, p. 281.

15. *Impact of Allied Air Interdiction on German Strategy for Normandy,* p. 1.

16. Fuller, *Decisive Battles, Vol. 3,* p. 560.

17. *Impact of Allied Air Interdiction on German Strategy for Normandy*, pp. 11–14
18. *USAF Tactical Operations: World War II and Korea*, p. 30.
19 Murray, p. 274, and Uebe, p. 100.
20. James, pp. 292–97.
21 Kenney, p. 275.
22. *The Yom Kippur War*, pp. 182–83.

CHAPTER 7

1. Eube, p. 25.
2. Cutler, pp. 99–100.
3 Oleg Hoeffding, *German Air Attacks Against Industry and Railroads in Russia, 1941–45* (Santa Monica, Calif.: Rand Corporation, 1970), pp. 25–27.
4. A. Goutard, *The Battle of France 1940*, transl. by A. R. P. Burgess (New York. Ives Washburn, 1959), p. 132.
5. Murray, p. 125.
6. Ibid., p. 282.
7. *Air Campaigns in the Pacific War*, p. 28.
8. Churchill, pp. 181–86.
9. Bekker, p. 195.
10. The Battle of Kursk in July–August 1943 was one of the largest tank battles of World War II. Kursk is an important agricultural center in the western Russian Soviet Federated Socialist Republic on the upper Seym River.
11. Bekker, pp. 434–35.
12. *Condensed Analysis of the 9th Air Force in the European Theater of Operations* (Washington, DC. US Army Air Forces, Office of Assistant Chief of Air Staff, Office of Air Force History, 1984, reprinted from 1946 edition), p. 29.
13. James, pp. 485–88.
14. Ibid., pp. 392–94.
15. *Air Campaigns in the Pacific War*, p. 32.
16. Bekker, p. 411.
17. Ibid., pp. 411, 421.
18. Ibid., p. 439.
19. Momyer, pp. 307–11.
20. Murray, p. 86.
21. Kenney, pp. 270–71.

CHAPTER 8

1. Arthur Bryant, *The Turn of the Tide* (New York. Doubleday & Company, Inc., 1957), p. 81.
2. Taylor, p. 164.
3. Ibid., p. 71.
4. Suchenwirth, pp 64–65.
5. Taylor, pp. 99, 164; and Bekker, pp. 243–46.
6. Bekker, pp 200–201, 223, 229, and Taylor, pp. 135, 138–39.
7. Taylor, pp. 151–59.
8. Bekker, p. 243, and Taylor, p. 164.
9. Taylor, pp. 163–65, and Bekker, pp. 247–48.
10 Galland, pp. 151–53.
11. Murray, p. 278.

12. Galland, pp 240–41.

13. Suchenwirth, p. 118.

14. Ibid., pp. 89–90.

CHAPTER 9

1. Erich von Manstein, *Lost Victories,* transl. and ed. by Anthony G. Powell (Novato, Calif.: Presidio Press, 1984), p 177.

2. Hoeffding, pp. 22–23.

3. Fuller, *Decisive Battles of the Western World, Vol 3*, p. 296.

4. Vicksburg held a strategic location on the Mississippi River, halfway between Memphis and New Orleans. It was beseiged for 47 days during Grant's campaign for control of the Mississippi River, before surrendering on 4 July 1863.

5. Russel F. Weigley, *The American Way of War* (Bloomington. Indiana University Press, 1977), pp. 317–21.

6. Wesley F. Craven and James L. Cates, eds., *The Army Air Forces in World War II, Vol. IV. The Pacific—Guadalcanal to Saipan (August 1943 to July 1944)* (Chicago. The University of Chicago Press, 1950), p. iv.

7. With the exception of the loss of Burma, Japanese lines in the China-Burma-India theater were not much different in August 1945 than in August 1942. *Summaries of Selected Military Campaigns.* (West Point, N.Y.: Department of Military Art and Engineering, US Military Academy, nd, Special Printing for Department of History, US Air Force Academy, 1960), pp. 152, 163.

8. *Air Campaigns of the Pacific War,* pp. 3, 4.

9. D. Clayton James, *The Years of MacArthur 1941–45* (Boston. Houghton Mifflin Company, 1975), p. 190.

10. Ibid., pp. 331–32.

11. Ibid.

12. Ibid., pp. 334–35.

13. *Air Campaigns of the Pacific War,* pp. 39–42.

14. James, pp. 607–9.

15. Martin van Creveld, *Supplying War: Logistics from Walenstein to Patton* (New York. Cambridge University Press, 1980), pp. 213–16.

16. Craven and Cates, *Army Air Forces in World War II, Vol. II,* pp. 428–30.

CHAPTER 10

1. The Israeli attack on Arab air in the June 1967 war is perhaps the only example of a single battle on a single day winning theater air superiority.

2. Isareli commandos rescued 91 passengers and 12 crew members of an Air France plane at Entebbe Airfield, Uganda, on 3 July 1976 in what has been called one of the most spectacularly successful rescue raids of modern times.

3. Richard Lee Scott, Jr., *Flying Tiger. Chennault of China* (Garden City, N.Y.. Doubleday & Company, Inc., 1959), pp. 70, 90–92.

4. Brown, p. 10.

5 Churchill, p. 82.

6. Bekker, pp. 430–31.

THE AIR CAMPAIGN IN RETROSPECT

1. Militiades (The Younger) was the general who led Athenian forces to victory over the Persians in the Battle of Marathon, in northeast Attica, in September 490

BC. At dawn, the Athenians advanced within a mile of the enemy by felling trees and making obstacles against the dreaded Persian cavalry. With a thin center and strengthened wings, the line of 10,000 Athenians and 1,000 Plataeans charged the enemy infantry before the Persian cavalry could return. The Greek wings defeated the Persians and wheeled inward to rout the Persian center, which had driven the Greek center back. The longer spears and heavier armor of the bronze-clad Greek infantrymen prevailed over the short spears, wicker shields, and padded clothing of the Persians.

Select Bibliography

Bekker, Cajus. *The Luftwaffe War Diaries*. Translated by Frank Ziegler. New York: Ballantine Books, 1969.

Brown, Anthony Cave. *Bodyguard of Lies*. New York: Bantam Books, Inc., 1976.

Bryant, Sir Arthur. *The Turn of the Tide*. New York: Doubleday & Company, Inc., 1957.

Churchill, Randolph S. and Winston S. *The Six Day War*. Boston: Houghton Mifflin Company, 1967.

Clausewitz, Carl von. *On War*. Translated and edited by Michel Howard and Peter Paret. Princeton, N.J.: Princeton University Press, 1976.

Cordesman, Anthony H. "The Sixth Arab-Israeli Conflict: Military Lessons for American Defense Planning." *Armed Forces Journal International*, August 1982.

Craven, Wesley F. and Cate, James L., Editors. *The Army Air Forces in World War II*. Volumes II & IV. Chicago: The University of Chicago Press, 1949 & 1950.

Cutler, Paul S. "ELTA Plays a Decisive Role in the EOB Scenario." *Military Electronics/Countermeasures*, January 1983.

———. "EW Won the Bekaa Valley Air Battle." *Military Electronics/Countermeasures*, January 1983.

———. "Lt. Gen. Rafael Eitan: We Learned Both Tactical and Technical Lessons in Lebanon." *Military Electronics/Countermeasures*, February 1983.

Department of Military Art and Engineering. *Summaries Of Selected Mlitary Campaigns*. West Point, N.Y.: US Military Academy, nd. Special Printing for Department of History, US Air Force Academy, 1960.

Dupuy, Colonel Trevor N. *A Genius for War: the German Army and General Staff, 1807–1945*. London: MacDonald and Jane's, 1977.

Fuller, J.F.C. *Decisive Battles of the Western World*. Volumes II & III. London: Eyre & Spottiswoode, 1963.

170

————. *The Generalship of Alexander the Great*. London: Eyre & Spottis-woode, 1958.

Futrell, Robert. F. *The United States Air Forces in Korea, 1950–53*. Washington, DC: Office of Air Force History, 1983. Revised Edition.

Galland, General Adolf. *The First and the Last*. Translated by Merwyn Savill. New York: Ballantine Books, 1963.

Goutard, Colonel A. *The Battle of France 1940*. Translated by A.R.P. Burgess. New York: Ives Washburn, 1959.

Gurney, Gene. *Five Down and Glory*. New York: Ballantine Books, 1965.

Hansell, Haywood S., Jr. *Strategic Air War Against Japan*. Maxwell AFB, Ala.: Airpower Research Institute, 1980.

Hoeffding, Oleg. *German Air Attacks Against Industry and Railroads in Russia, 1941–1945*. Santa Monica, Calif.: Rand Corporation, 1970.

Irving, David. *The Trail of the Fox*. New York: Avon Books, 1978.

James, D. Clayton. *The Years of MacArthur 1941–1945*. Boston: Houghton Mifflin Company, 1975.

Kenney, George C. *General Kenney Reports*. New York: Duell, Sloan, and Pearce, 1949.

Lambeth, Benjamin S. *Moscow's Lessons from the 1982 Lebanon Air War*. Santa Monica, Calif.: Rand Corporation, 1984.

Lavalle, A.J.C., Editor. *The Tale of Two Bridges and the Battle for the Skies over North Vietnam*. US Air Force Southeast Asia Monograph Series. Washington, DC: Superintendent of Documents, 1976.

————. *The Vietnamese Air Force 1951–1975: An Analysis of Its Role in Combat*. US Air Force Southeast Asia Monograph Series, Vol. 3, Monographs 4–5, 1977. Washington, DC: Superintendent of Documents, 1976.

Lewin, Ronald. *Rommel: As Military Commander*. New York: Ballantine Books, 1972.

Manchester, William. *American Caesar: Douglas MacArthur 1880–1964*. Boston: Little, Brown and Company, 1978.

Momyer, William W. *Air Power in Three Wars (WWII, Korea, Vietnam)*. Washington, DC: US Air Force, 1978.

Morrow, John H. *German Air Power in World War I*. Lincoln: University of Nebraska Press, 1982.

Murray, Williamson, *Strategy for Defeat: The Luftwaffe 1933–45*. Maxwell AFB, Ala.: Air University Press, 1983.

Plocher, Hermann. *The German Air Force Versus Russia, 1941*. Maxwell AFB, Ala.: Air University Press, 1965.

Prange, Gordon W. *Miracle at Midway*. New York: McGraw-Hill Book Company (Penguin Books), 1983.

Robinson, Clarence A. Jr. "Surveillance Integration Pivotal in Israeli Success." *Aviation Week and Space Technology*. 5 July 1982.

Salagar, F.M. *Operation "Strangle" (Italy, Spring 1944): A Case Study of Tactical Air Interdiction*. Santa Monica, Calif.: Rand Corporation, 1972.

Schwabedissen, Walter. *The Russian Air Force in the Eyes of German Commanders.* Maxwell AFB, Ala.: Air University Press, 1960.

Scott, Robert Lee, Jr. *Flying Tiger: Chennault of China.* Garden City, N.Y.: Doubleday & Company, Inc., 1959.

Speer, Albert. *Inside the Third Reich.* Translated by Richard and Clara Winston. New York: Avon Books, 1971.

Suchenwirth, Richard. *Historical Turning Points in the German Air Force War Effort.* Maxwell AFB, Ala.: Air University Research Studies Institute, 1959.

The Insight Team of the London *Sunday Times. The Yom Kippur War.* Garden City, N.Y.: Doubleday & Company, Inc., 1974.

Taylor, Telford. *The Breaking Wave.* New York: Simon and Schuster, 1967.

Uebe, D. Klaus. *Russian Reactions to German Airpower in World War II.* Maxwell AFB, Ala.: Aerospace Studies Institute, 1964.

US Air Force Assistant Chief of Staff, Studies and Analysis. *The Impact of Allied Air Interdiction on German Strategy for Normandy.* Washington, DC: Headquarters, US Air Force, 1969.

————. *The Relationship Between Sortie Ratios and Loss Rates for Air-to-Air Battle Engagements During World War II and Korea.* Washington, DC: Headquarters, US Air Force, 1970. US Air Force Historical Division Liaison Office. *USAF Tactical Operations: World War II and Korea.* Washington, DC: US Air Force, 1962.

US Army Air Forces, Office of Assistant Chief of Air Staff. *Condensed Analysis of the Ninth Air Force in the European Theater of Operations.* Washington, DC: Headquarters, Army Air Forces, Office of Assistant Chief of Air Staff, 1946. Reprinted Washington, DC: Office of Air Force History, 1984.

US Strategic Bombing Survey. *Air Campaigns in the Pacific War.* Washington, DC: US Government Printing Office, 1947.

————. *The Campaigns of the Pacific War.* Washington, DC: US Government Printing Office, 1946.

————. *Japanese Air Power.* Washington, DC: Military Analysis Division, 1946.

————. *Overall Report (European War).* Washington, DC: US Government Printing Office, 1945.

————. *Summary Report (Pacific War).* Washington, DC: US Government Printing Office, 1946.

Van Creveld, Martin. *Supplying War: Logistics from Wallenstein to Patton.* New York: Cambridge University Press, 1980.

Von Manstein, Field Marshal Erich. *Lost Victories.* Edited and translated by Anthony G. Powell. Novato, Calif.: Presidio Press, 1984.

Weigley, Russel F. *The American Way of War.* Bloomington: Indiana University Press, 1977.

Weizman, Major General Ezer. *On Eagles' Wings.* New York: MacMillan Publishing Co., Inc., 1976.

Index

The Author

John A. Warden III is an executive, strategist, planner, author, fighter pilot, and motivational speaker with a worldwide reputation for innovations in military, political, educational, and commercial endeavors. He retired as a colonel from the United States Air Force in 1995 and started his own company, Venturist, Incorporated, a firm that specializes in corporate strategy development, innovation, and multi-media development.

John Warden wrote *The Air Campaign: Planning For Combat* in 1986 while attending the National War College. During his career, he developed a method to simplify analysis of complex organizations, articulated the radically new concept of parallel war, devised a new approach to education, and synthesized a new approach to business strategy. In addition to being a theorist, however, he has had major successes in the real world of war, politics, education, and business: Generals Schwarzkopf and Powell credited him with creating the air campaign that defeated Iraq in the Gulf War; while serving as Special Assistant to the Vice President of the United States, he led successful efforts to remove government obstacles to American competitiveness, championed reforms to allow American defense firms to compete more effectively in the commercial sector, and co-authored a major manufacturing agreement between Japan and the United States; as head of the Air Force's Air Command and Staff College, he developed and put into effect entirely new educational concepts and made the College a model for institutions of higher learning in the United States and abroad; and as a fighter pilot and commander of the 36th Tactical Fighter Wing in Bitburg, Germany, he introduced new aerial tactics and command and control methodology. He has over 3,000 flying hours in aircraft such as the F-15, F-4, and OV-10. He flew 266 combat missions as a forward air controller over Vietnam and Laos in the Vietnam War. His decorations include the Distinguished Service Medal, the Defense Superior Service Medal, the Legion of Merit, the Distinguished Flying Cross, and the Air Medal with ten Oak Leaf Clusters.

John Warden has appeared on ABC, CBS, CNN, PBS, BBC, the History Channel, and the Discovery Channel and is frequently featured or cited in newspapers and magazines with global circulation. He is currently writing a book on business strategy due out the summer of 1999. He has a BS from the United States Air Force Academy, an MA from Texas Tech University, and is a graduate of the National War College. He is originally from McKinney, Texas, and now makes his home in Montgomery, Alabama. He is married and has two children; his daughter is Vice President of Venturist, Inc. and his son is flying the most potent military machine in the world – the B-2 stealth bomber.

For more information about Venturist, Inc. or to contact John Warden, please visit www.venturist.com or email jwarden@venturist.com.

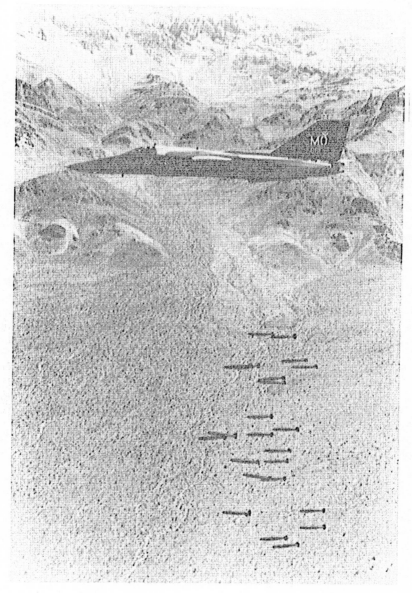

Training for the Air Campaign: An F-111 airdrops 24 Mark-82 low-drag bombs over the Nellis AFB range in Nevada.